HOLT

Physics

Section Quizzes

HOLT, RINEHART AND WINSTON

A Harcourt Education Company

Orlando • **Austin** • New York • San Diego • Toronto • London

ISBN 0-03-036836-7

6 7 8 9 018 09 08 07

Contents

The Science of Physics

What Is Physics? 1
Measurements in Experiments 3
The Language of Physics 5

Motion in One Dimension

Displacement and Velocity 7
Acceleration 9
Falling Objects 11

Two-Dimensional Motion and Vectors

Introduction to Vectors 13
Vector Operations 15
Projectile Motion 17
Relative Motion 19

Forces and the Laws of Motion

Changes in Motion 21
Newton's First Law 23
Newton's Second and Third Laws 25
Everyday Forces 27

Work and Energy

Work 29
Energy 31
Conservation of Energy 33
Power 35

Momentum and Collisions

Momentum and Impulse 37
Conservation of Momentum 39
Elastic and Inelastic Collisions 41

Circular Motion and Gravitation

Circular Motion 43
Newton's Law of Universal Gravitation 45
Motion in Space 47
Torque and Simple Machines 49

Fluid Mechanics

Fluids and Buoyant Force 51
Fluid Pressure 53
Fluids in Motion 55

Heat

Temperature and Thermal Equilibrium 57
Defining Heat 59
Changes in Temperature and Phase 61

Thermodynamics

Relationships Between Heat and Work 63
The First Law of Thermodynamics 65
The Second Law of Thermodynamics 67

Vibrations and Waves

Simple Harmonic Motion 69
Measuring Simple Harmonic Motion 71
Properties of Waves 73
Wave Interactions 75

Sound

Sound Waves 77
Sound Intensity and Resonance 79
Harmonics 81

Light and Reflection

Characteristics of Light 83
Flat Mirrors 85
Curved Mirrors 87
Color and Polarization 89

Refraction

Refraction 91
Thin Lenses 93
Optical Phenomena 95

Interference and Diffraction

Interference 97
Diffraction 99
Lasers 101

Electric Forces and Fields

Electric Charge 103
Electric Force 105
The Electric Field 107

Electrical Energy and Current

Electric Potential 109
Capacitance 111
Current and Resistance 113
Electric Power 115

Circuits and Circuit Elements

Schematic Diagrams and Circuits 117
Resistors in Series or in Parallel 119
Complex Resistor Combinations 121

Magnetism

Magnets and Magnetic Fields 123
Magnetism from Electricity 125
Magnetic Force 127

Electromagnetic Induction

Electricity from Magnetism 129
Generators, Motors, and Mutual Inductance 131
AC Circuits and Transformers 133
Electromagnetic Waves 135

Atomic Physics

Quantization of Energy 137
Models of the Atom 139
Quantum Mechanics 141

Subatomic Physics

The Nucleus 143
Nuclear Decay 145
Nuclear Reactions 147
Particle Physics 149

The Science of Physics

Section Quiz: What Is Physics?

Write the letter of the correct answer in the space provided.

_____ **1.** Physics is concerned with which of the following matters?
  a. applying physics principles
  b. describing the physical world
  c. making predictions about a broad range of phenomena
  d. all of the above

_____ **2.** Measuring temperature would most likely involve which major area in physics?
  a. optics
  b. thermodynamics
  c. relativity
  d. vibrations and wave phenomena

_____ **3.** A physicist who studies the behavior of submicroscopic particles is working in which area within physics?
  a. mechanics
  b. electromagnetism
  c. relativity
  d. quantum mechanics

_____ **4.** After making observations and collecting data that leads to a question, a physicist will then
  a. formulate and test hypotheses by experimentation.
  b. state conclusions.
  c. interpret the results.
  d. revise the initial hypotheses.

_____ **5.** How many variables may be tested legitimately in any one experiment?
  a. as many as a physicist can handle
  b. three
  c. one
  d. five

_____ **6.** Why do physicists use models?
  a. They are usually easy to build.
  b. They are helpful when explaining fundamental features.
  c. They are usually inexpensive.
  d. none of the above

The Science of Physics *continued*

_____ **7.** Which of the following represents a system?
  a. flag blowing in the wind
  b. ball rolling on the ground
  c. picture hanging on the wall
  d. all of the above

_____ **8.** Models can do all of the following *except*
  a. explain every aspect of natural phenomena.
  b. help build hypotheses.
  c. guide experimental design.
  d. make predictions in new situations.

9. Describe a situation that uses the processes of the scientific method. Explain how the scientific processes are used in that situation.

_____

_____

_____

_____

_____

_____

10. What is the first step toward simplifying a complicated situation, and what are three ways to summarize a complicated situation?

_____

_____

_____

_____

_____

Assessment

The Science of Physics

Section Quiz: Measurements in Experiments

Write the letter of the correct answer in the space provided.

_____ **1.** What is the SI base unit for length?
  a. meter
  b. kilogram
  c. kilometer
  d. second

_____ **2.** What quantity does the kilogram measure?
  a. time
  b. distance
  c. force
  d. mass

_____ **3.** In scientific notation, 674.3 mm equals
  a. 0.6743×10^{-3} mm.
  b. 6.743×10^{3} km.
  c. 6.743×10^{2} mm.
  d. 6.743×10^{2} m.

_____ **4.** In scientific notation, 0.000 005 823 μg equals
  a. 5.823×10^{-6} μg.
  b. 5.823×10^{-12} g.
  c. 5.823×10^{-9} mg.
  d. all of the above

_____ **5.** The average mass of a proton is 1.673×10^{-27} kg. What is this mass in grams?
  a. 1.673×10^{-30} g
  b. 1.673×10^{-24} g
  c. 1.673×10^{-27} g
  d. 1.673×10^{-81} g

_____ **6.** The accepted value for free-fall acceleration is 9.806 65 m/s². Which of the following measurements is the most accurate?
  a. 9.808 60 m/s²
  b. 9.906 65 m/s²
  c. 8.806 77 m/s²
  d. 9.006 65 m/s²

_____ **7.** Precision describes
    a. human error.
    b. the relationship of a measurement to an accepted standard.
    c. the limitations of the measuring instrument.
    d. the lack of instrument calibration.

_____ **8.** How many significant figures does 50.003 00 have?
    a. five
    b. seven
    c. two
    d. three

9. How do significant figures indicate a measurement's precision?

_____

_____

_____

_____

10. Calculate the area of a room whose length is 15.23 m and width is 8.7 m. Express your answer in scientific notation and with the correct number of significant digits.

Name _____ Class _____ Date _____

The Science of Physics

Section Quiz: The Language of Physics

Write the letter of the correct answer in the space provided.

TABLE 1 DATA FROM HEATING EXPERIMENT

Time (s)	Substance A Temp (°C)	Substance B Temp (°C)
0.0	22.8	22.8
10.0	25.9	23.5
20.0	32.4	24.1
30.0	45.1	24.9

_____ **1.** Based on the data from Table 1, which of the following statements is correct?
   a. The temperature increased equally during each time period for both substances.
   b. There is no relationship between heating time and temperature for either substance.
   c. As time increased, the temperature increased for both substances.
   d. none of the above

GRAPH 1 DATA FROM HEATING EXPERIMENT

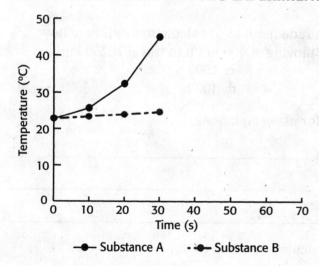

_____ **2.** What does Graph 1 show about the heating rate of substance A versus substance B?
   a. Compared to substance B, substance A has a faster heating rate.
   b. Compared to substance A, substance B has a slower heating rate.
   c. Substance A and B heat at different rates.
   d. all of the above

The Science of Physics *continued*

_____ **3.** Which of the following equations best shows the average relationship between temperature and time for substance B as given in Table 1 and Graph 1?

a. $\Delta T = 0.07(\Delta t)$

b. $\Delta T = 0.07(\Delta t)^2$

c. $(\Delta T)^2 = 0.7(\ t)$

d. $\Delta T = 7.4(\Delta t)$

_____ **4.** What does the symbol m represent?

a. change in mass

b. change in meters

c. difference in motion

d. distance in meters

_____ **5.** What is the standard abbreviation for seconds?

a. sec

b. s

c. sds

d. t

_____ **6.** All of the following are unit abbreviations *except* which one?

a. m

b. kg

c. y

d. s

_____ **7.** If the final answer's dimension is to be in length, which of the following operations is correct?

a. (time/length) × time

b. time × (length/time)

c. (time × length) − length

d. length × (length/time)

_____ **8.** Using the order-of-magnitude method of calculation, estimate how long it would take a car moving at 109 km/h to travel 10450 km.

a. 100 000 h

b. 10 000 h

c. 1000 h

d. 100 h

9. Name at least two advantages for using equations.

_____

_____

_____

10. What are order-of-magnitude calculations used to do?

_____

_____

_____

_____

Name _____ Class _____ Date _____

Motion in One Dimension

Section Quiz: Displacement and Velocity

Write the letter of the correct answer in the space provided.

_____ **1.** Which of the following situations represents a positive displacement of a carton? Assume positive position is measured vertically upward along a y-axis.
  a. A delivery person waiting for an elevator lowers a carton onto a dolly.
  b. When the elevator doors open, the delivery person lifts the dolly over the threshold of the elevator.
  c. The delivery person pushes the dolly to the back of the elevator while pressing a floor button.
  d. The door closes and the elevator moves from the 10th to the 4th floors.

Refer to the figure below to answer questions 2–4.

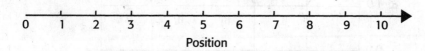

Position

_____ **2.** What is the correct description of any change of position farther to the right of zero?
  a. positive displacement
  b. positive distance
  c. positive position
  d. positive change of displacement

_____ **3.** A dog walks from +4 m to +2 m. Which of the following statements is true about the dog's motion?
  a. $x_i = +2$ m
  b. $x_f = +2$ m
  c. $\Delta x = +2$ m
  d. $v_{avg} = 2$ m/s

_____ **4.** What is the maximum negative displacement a dog could have if it started its motion at +3 m?
  a. +7 m
  b. +3 m
  c. −3 m
  d. −7 m

_____ **5.** Rank in decreasing order the displacements of objects having the following pairs of average velocity and time of motion.
  I. $v_{avg} = +2.0$ m/s, $\Delta t = 2.0$ s
  II. $v_{avg} = +3.0$ m/s, $\Delta t = 2.0$ s
  III. $v_{avg} = -3.0$ m/s, $\Delta t = 3.0$ s
  a. I, II, III
  b. II, III, I
  c. II, I, III
  d. III, II, I

_____ **6.** Rank in decreasing order the distances traveled by objects having the
following pairs of average velocity and time of motion.
I. $v_{avg} = +2.0$ m/s; $\Delta t = 2.0$ s
II. $v_{avg} = +3.0$ m/s, $\Delta t = 2.0$ s
III. $v_{avg} = -3.0$ m/s, $\Delta t = 3.0$ s

  a. I, II, III   **c.** II, I, III
  b. II, III, I   **d.** III, II, I

The graph below shows the motion of a dog pacing along a fence. Refer to the graph to answer questions 7–10.

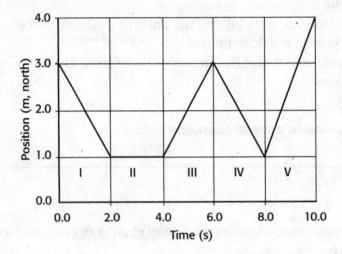

_____ **7.** For the five time intervals shown, during how many intervals does the
dog have the same average velocity?

  a. 0   **c.** 2
  b. 1   **d.** 3

_____ **8.** For the five time intervals shown, during how many intervals does the
dog pace at the same average speed?

  a. 0   **c.** 2
  b. 1   **d.** 3

9. Describe the dog's motion when it is at 1.0 m.

_____

_____

10. What is the dog's average velocity for total displacement?

Motion in One Dimension

Section Quiz: Acceleration

Write the letter of the correct answer in the space provided.

_____ **1.** The average acceleration is the ratio of which of the following quantities?
a. $\Delta d{:}\Delta v$
b. $d{:}\Delta t$
c. $v{:}\Delta v$
d. $\Delta v{:}\Delta t$

_____ **2.** The speed of a car will increase if the car's
a. initial velocity is positive and its acceleration is zero.
b. initial velocity is positive and its acceleration is positive.
c. initial velocity is positive and its acceleration is negative.
d. initial velocity is negative and its acceleration is positive.

_____ **3.** For a scooter with a negative acceleration, which of the following statements is always true?
a. The scooter is losing speed.
b. The final velocity of the scooter will be negative.
c. The initial velocity of the scooter will be greater than its final velocity.
d. The scooter will have a negative displacement.

Questions 4–9 refer to the following velocity-time graph of a jogger. The positive direction is away from the jogger's home.

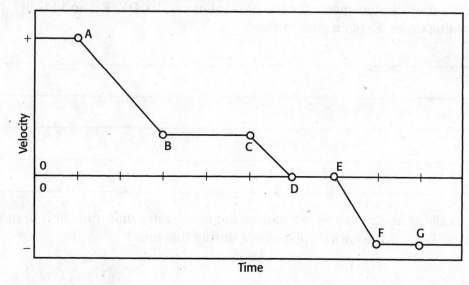

_____ **4.** The jogger is at rest during which interval?
a. AB   **c.** DE
b. BC   **d.** EF

Motion in One Dimension *continued*

_____ **5.** During which of the following intervals does the jogger have a constant positive velocity?

    a. BC

    b. CD

    c. DE

    d. FG

_____ **6.** During which interval is the magnitude of the jogger's acceleration the greatest?

    a. AB

    b. BC

    c. DE

    d. EF

_____ **7.** During which of the following intervals is the speed of the jogger decreasing?

    a. AB only

    b. AB and CD

    c. AB, CD, and EF

    d. EF only

_____ **8.** During which of the following intervals is the jogger's motion toward home?

    a. CD and EF

    b. DE only

    c. EF only

    d. EF and FG

9. Rank the four displacements for the time intervals CD, DE, EF, and FG in decreasing order. Explain your answer.

_____

_____

_____

_____

_____

10. A cat walking at 0.25 m/s sees a mouse and accelerates uniformly at 0.40 m/s^2 for 3.0 s. What is the cat's displacement during this time?

Name _____ Class _____ Date _____

Motion in One Dimension

Section Quiz: Falling Objects
Write the letter of the correct answer in the space provided.

_____ **1.** An object in free fall
    a. experiences no air resistance.
    b. undergoes a downward acceleration.
    c. has an acceleration with a magnitude of 9.81 m/s^2 near Earth's surface.
    d. all of the above

_____ **2.** An object in free fall
    a. experiences an increase in speed of 9.81 m/s during each second.
    b. moves only downward.
    c. undergoes a velocity decrease of 9.81 m/s during each second.
    d. all of the above

_____ **3.** The displacement of an object undergoing free fall from rest is proportional to
    a. $\frac{1}{2}\Delta t.$ **c.** $(\Delta t)^2.$
    b. $\Delta t.$ **d.** $(2t)^2.$

_____ **4.** The final velocity of an object undergoing free fall from rest is proportional to
    a. $\frac{1}{2}\Delta x.$ **c.** $\sqrt{(\Delta x)}.$
    b. $\Delta x.$ **d.** $(\Delta x)^2.$

_____ **5.** The graph below shows the motion of four objects. Which of the following lines represent an object in free fall? Assume positive velocity is upward.

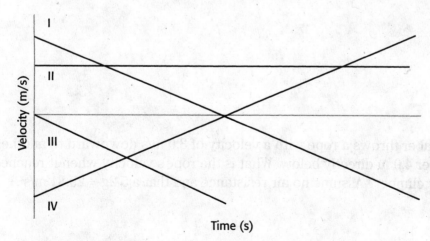

    a. I only   **c.** I and III
    b. II only   **d.** I and IV

| **Motion in One Dimension** *continued*

Questions 6–8 refer to the following demonstration.

A red ball is dropped from rest and undergoes free fall. One second later a blue ball is dropped from rest and undergoes free fall.

_____ **6.** The red ball's change of velocity during the third second of the demonstration is Δv_{2-3}. What is the change of velocity of the blue ball during this time interval?

a. $\frac{1}{2}\Delta v_{2-3}$

b. Δv_{2-3}

c. $2\Delta v_{2-3}$

d. $3\Delta v_{2-3}$

_____ **7.** The red ball's velocity at the end of 3.0 seconds is v_3. What is the blue ball's velocity at this moment of the demonstration?

a. $\frac{1}{9}v_3$ c. $\frac{1}{2}v_3$

b. $\frac{1}{4}v_3$ d. $\frac{2}{3}v_3$

_____ **8.** The displacement of the blue ball during the time interval 3.0–4.0 s is equal to the displacement of the red ball in which of the following intervals of the demonstration?

a. 0.0–1.0 s

b. 1.0–2.0 s

c. 2.0–3.0 s

d. 3.0–4.0 s

9. A juggler throws a ball vertically upward well above his reach. Sketch a graph of the ball's speed versus time. Show the motion of the ball from the moment it leaves the juggler's hand to the moment the juggler catches it. Assume no air resistance and that the ball is in free fall.

10. A climber throws a rope with a velocity of 3.0 m/s downward to another climber 4.0 m directly below. What is the rope's velocity when it reached the lower climber? Assume no air resistance and that a 5 2g = 29.81 m/s².

Name _____ Class _____ Date _____

Two-Dimensional Motion and Vectors

Section Quiz: Introduction to Vectors

Write the letter of the correct answer in the space provided.

_____ 1. In a diagram, the length of a vector arrow represents the
     a. type of vector.
     b. direction of the vector.
     c. magnitude of the vector.
     d. cause of the vector.

_____ 2. Which of the following quantities used to describe motion is an example of a vector quantity?
     a. distance
     b. speed
     c. time
     d. average velocity

_____ 3. A vector remains unchanged
     a. if it is moved in any direction.
     b. only if it is moved parallel to its original direction.
     c. only if it is rotated perpendicular to its original direction.
     d. only if it is *not* moved.

Refer to the figure below to answer questions 4–6.

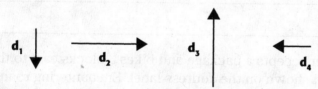

_____ 4. Which of the displacements, when added in the order given, will yield a displacement equal to $d_1 + d_2 + d_3$?
     a. $d_3 + d_4 + d_1$
     b. $d_2 + d_1 + d_3$
     c. $d_2 + d_4 + d_3$
     d. $d_4 + d_3 + d_1$

_____ 5. Which of the following represents the vector resulting in the subtraction of the displacement vectors $d_2 - d_1$ shown in the figure?

Two-Dimensional Motion and Vectors *continued*

_____ **6.** In the figure,
   a. $d_1 = d_4$.
   b. $d_1 = -d_4$.
   c. $d_2 = -2d_4$.
   d. $d_3 = \frac{1}{2}d_1$.

_____ **7.** If vectors are moved according to the rules of the triangle method of vector addition, then the resultant vector is drawn
   a. from the tail of the first vector to the tail of the last vector.
   b. from the tail of the first vector to the tip of the last vector.
   c. from the tip of the first vector to the tail of the last vector.
   d. from the tip of the first vector to the tip of the last vector.

_____ **8.** A skier slides onto a horizontal patch of slushy snow at velocity v_{skier}, and slows to a constant speed without changing direction. Which of the following expressions could be the skier's resulting velocity?
   a. $-3v_{skier}$
   b. $+\frac{1}{4}v_{skier}$
   c. $-\frac{1}{5}v_{skier}$
   d. all of the above

9. What is the major difference between a vector quantity and a scalar quantity?

_____

_____

_____

10. A bicycle courier accepts a package and bikes 3 blocks east to the intersection of the street shown on the address label. Encountering road construction, the courier detours by continuing one block east, one block north, and one block west. The courier then bikes two blocks north to the address. What is the resultant displacement of the courier? Use the graphical method.

Assessment

Two-Dimensional Motion and Vectors

Section Quiz: Vector Operations

Write the letter of the correct answer in the space provided.

_____ **1.** In a diagram, a student draws a vector representing the velocity of a plane traveling at 525 km/h 18.0° east of north as a vector lying along the +x-axis. To represent the velocity of another plane traveling at 478 km/h 75.0° east of north on the diagram, at what orientation with respect to the +x-axis should the student draw this vector?

    a. 93.0° clockwise from the +x-axis
    b. 75.0° clockwise from the +x-axis
    c. 57.0° clockwise from the +x-axis
    d. 75.0° counterclockwise from the +x-axis

_____ **2.** In question 1, another student chose east as the direction of the +x-axis. At what orientation with respect to the +x-axis should this student draw the vector representing the velocity of the second plane?

    a. parallel to the +x-axis
    b. 90.0° clockwise from the +x-axis
    c. 75.0° clockwise from the +x-axis
    d. 15.0° counterclockwise from the +x-axis

_____ **3.** In a school playground, a child runs 5 m in the x-direction and then 2.0 m in the $-y$-direction. Which of the following expressions represents the magnitude of the child's resultant displacement?

    a. $(5 \text{ m}) + (-2 \text{ m})$ **c.** $\sqrt{(5 \text{ m})^2 - (2 \text{ m})^2}$
    b. $\sqrt{(5 \text{ m})} - \sqrt{(2 \text{ m})}$ **d.** $\sqrt{(5 \text{ m})^2 + (-2 \text{ m})^2}$

_____ **4.** In the triangle below, what does θ equal?

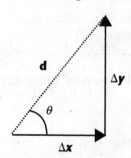

    a. $\tan^{-1}\left(\dfrac{\Delta y}{\Delta x}\right)$ **c.** $\tan^{-1}\left(\dfrac{\Delta x}{d}\right)$

    b. $\tan^{-1}\left(\dfrac{\Delta x}{\Delta y}\right)$ **d.** $\tan^{-1}\left(\dfrac{d}{\Delta y}\right)$

| Two-Dimensional Motion and Vectors *continued*

_____ **5.** The projection of a vector along the axes of a coordinate system is
called
  a. a component of the vector.
  b. a tangent of the vector.
  c. the resultant of the vector.
  d. the magnitude of the vector.

_____ **6.** The components of a vector are
  a. each equal to half the magnitude of the vector.
  b. independent of the orientation of the vector.
  c. perpendicular.
  d. vector quantities.

_____ **7.** A vector that has components that lie along the $-x$-axis and the
$+y$-axis is oriented at angle A measured counterclockwise from the
$+x$-axis. Which of the following ranges give the values of angle A?
  a. $0° < A < 90°$
  b. $90° < A < 180°$
  c. $180° < A < 270°$
  d. $270° < A < 360°$

_____ **8.** If the x-component of **B** equals the magnitude of **B**, then
  a. both components are equal.
  b. B_y equals $\sqrt{B^2 + B_x^2}$.
  c. B_y equals zero.
  d. B_y lies along the $-y$-axis.

9. What must you do to non-perpendicular vectors before you can use the
Pythagorean theorem to calculate the resultant of the vectors?

_____

_____

_____

10. In a blinding blizzard, a reindeer trudges 310 m 45° east of south across the
tundra. How far south does the reindeer move?

Assessment

Two-Dimensional Motion and Vectors

Section Quiz: Projectile Motion

Write the letter of the correct answer in the space provided.

_____ **1.** Which of the following may be classified as projectile motion?
    a. a punted football
    b. a thrown baseball
    c. a water droplet cascading down a waterfall
    d. all of the above

_____ **2.** In the absence of air resistance, the path of a projectile is a(n)
    a. parabola.
    b. arc.
    c. polygon.
    d. semicircle.

_____ **3.** The motion of a projectile in free fall is characterized by
    a. $a_x = a_y = -g$.
    b. $a_x = $ constant and $v_y = $ constant.
    c. $v_x = $ constant and $a_y = -g$.
    d. $v_x = $ constant and $v_y = $ constant.

_____ **4.** For an object to be a projectile, it must be in free fall and its initial velocity must
    a. have a horizontal component.
    b. have both a vertical and a horizontal component.
    c. have either a vertical or a horizontal component.
    d. start from rest.

_____ **5.** A baby drops a ball from her hand resting on the serving tray of her high chair. Simultaneously, she knocks another ball from the same tray. Which of the following statements are true?
    I. Both balls strike the ground at the same time.
    II. The dropped ball reaches the ground first.
    III. The knocked ball reaches the ground first.
    IV. Both balls strike the ground at the same speed.
    a. I only
    b. II only
    c. III only
    d. I and IV

Two-Dimensional Motion and Vectors *continued*

_____ **6.** A gardener holds the nozzle of a hose constant at a small angle above
the horizontal and observes the path of the stream of water coming
from the nozzle. If the pressure of the water is increased so that the
water leaves the nozzle at a greater speed,
  a. the height and width of the water's path will increase.
  b. the height of the water's path will increase but the width of the path
  will remain the same.
  c. the width of the water's path will increase but the height will remain
  the same.
  d. The height and width of the water's path will remain the same.

_____ **7.** Assuming no air friction and $a_y = -g$, the horizontal displacement of a
projectile depends on the
  a. horizontal component of its initial velocity only.
  b. vertical component of its initial velocity only.
  c. vertical component of its initial velocity and its time in flight.
  d. vertical component and the horizontal component of its initial
  velocity.

_____ **8.** A volleyball player taps a volleyball well above the net. The ball's
speed is least
  a. just after it is tapped by the player.
  b. at the highest point of its path.
  c. just before it strikes the ground.
  d. when the horizontal and vertical components of its velocity are
  equal.

9. Explain how a projectile can have a horizontal displacement even though its
vertical displacement is zero.

_____

_____

_____

10. In a movie production, a stunt person must leap from a balcony of one building
to a balcony 3.0 m lower on another building. If the buildings are 2.0 m apart,
what is the minimum horizontal velocity the stunt person must have to accom-
plish the jump? Assume no air resistance and that $a_y = -g = -9.81$ m/s^2.

Assessment

Two-Dimensional Motion and Vectors

Section Quiz: Relative Motion

Write the letter of the correct answer in the space provided.

_____ 1. Which of the following is a frame of reference for measuring motion?
    a. coordinate system
    b. diagram
    c. graph
    d. none of the above

Questions 2–6 refer to the following situation.

In a circus parade, a clown standing on a float moving at a constant forward speed drops a fake dumbbell. A child in the bleachers on the sidewalk observes the dumbbell. Ignore air resistance.

_____ 2. Who would observe the dumbbell falling straight down?
    a. the clown only
    b. the child only
    c. both the clown and the child
    d. neither the clown nor the child

_____ 3. Who would observe the dumbbell falling backward?
    a. the clown only
    b. the child only
    c. both the clown and the child
    d. neither the clown nor the child

_____ 4. The clown and the child would agree about which of the following observations?
    a. the time interval for the dumbbell to fall
    b. the vertical distance the dumbbell falls
    c. the initial vertical velocity of the dumbbell
    d. all of the above

_____ 5. The dumbbell's motion as described by an acrobat walking in the parade is very similar to the clown's description of the dumbbell's motion. What do you know about the motion of the acrobat?
    a. The acrobat is at rest with respect to the clown.
    b. The acrobat is at rest with respect to the float.
    c. both A and B
    d. neither A nor B

_____ **6.** If the float has a velocity of $+v_{float}$ with respect to the ground, what is the velocity of the clown as observed by the child?

a. 0

b. $+v_{float}$

c. $-v_{float}$

d. $\dfrac{v_{float} - v_{child}}{2}$

_____ **7.** The velocity of a passenger relative to a boat is $-v_{pb}$. The velocity of the boat relative to the river it is moving on is $+v_{br}$. The velocity of the river to the shore is $+v_{rs}$. What is the velocity of the passenger relative to the shore? Assume the + direction is east for all observers.

a. $-v_{pb} + v_{br} + v_{rs}$

b. $v_{pb} - (v_{br} + v_{rs})$

c. $-(v_{pb} + v_{br} + v_{rs})$

d. $v_{pb} + v_{br} + v_{rs}$

_____ **8.** A car traveling at the legal speed limit has velocity $+v_{car}$ with respect to the road. You are in another car traveling in the opposite direction at the legal speed limit with respect to the road. As the first car approaches you, what is its velocity relative to you?

a. $+v_{car}$

b. v_{car}

c. $+2v_{car}$

d. $-2v_{car}$

9. The figure below shows the motion, as seen by a stationary observer, of a stunt dummy (the black circle) as it dropped from a plane during the filming of a movie.

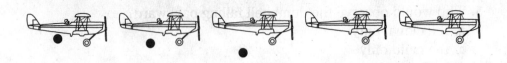

Another plane, flying at the same velocity and altitude as the first plane, carries a camera that is filming the drop. If the camera is accidentally dropped at the moment the dummy is dropped, describe the camera's recording of the dummy's motion during its fall.

_____

_____

10. A duck maintains a constant swimming speed of 0.50 m/s relative to the water. As it swims across a creek that has a constant downstream current of 0.30 m/s, a naturalist sitting on the bank observes its motion. What is the duck's speed as observed by the naturalist?

Name _____ Class _____ Date _____

Forces and the Laws of Motion

Section Quiz: Changes in Motion

Write the letter of the correct answer in the space provided.

_____ **1.** An action exerted on an object which may change the object's state of rest or motion defines
  a. acceleration.
  b. force.
  c. mass.
  d. velocity.

_____ **2.** Units that measure weight are units of
  a. acceleration.
  b. force.
  c. mass.
  d. velocity.

_____ **3.** Which of the following statements are true about the concept of field forces?

  I. Field forces can be exerted by objects that are in physical contact.
  II. Field forces can be exerted by objects that are not in contact.
  III. There are four fundamental field forces.

  a. I only
  b. II only
  c. II and III
  d. I, II, and III

_____ **4.** A force unit is defined as the product of a mass unit and a unit of
  a. time.
  b. displacement.
  c. velocity.
  d. acceleration.

_____ **5.** Which of the following statements describe force diagrams?

  I. Force diagrams show force vectors as arrows.
  II. Forces exerted on the object are represented.
  III. Forces exerted by the object are represented.
  a. I and II
  b. I and III
  c. II and III
  d. I, II, and III

_____ **6.** Which of the following statements describe free-body diagrams?

     I. Force diagrams show force vectors as arrows.

     II. Forces exerted on the object are represented.

     III. Forces exerted by the object are represented.

     a. I and II

     b. I and III

     c. II and III

     d. I, II, and III

_____ **7.** The gravitational force exerted on an object would most likely be represented by which force vector?

     a. ↑

     b. →

     c. ↓

     d. ←

_____ **8.** Two books are lying next to each other on a library table. The force due to gravity on the first book is 9 N and on the second is 13 N. The gravitational force on the table itself is 125 N. The floor supports the table with an upward force of 147 N. In a free-body diagram of the table, how many contact forces and how many field forces should be represented?

     a. 1, 2

     b. 1, 3

     c. 2, 2

     d. 3, 1

9. Sequence the steps in drawing a free-body diagram.

_____

_____

_____

10. A draft horse exerts a horizontal force of 600 N dragging a log, on which the force of friction is 500 N. The force due to gravity on the log is 700 N and the ground exerts an upward force of 700 N on it. Draw a free-body diagram of the log.

Assessment

Forces and the Laws of Motion

Section Quiz: Newton's First Law

Write the letter of the correct answer in the space provided.

_____ 1. Which of the following situations describes inertia?
   a. A stationary object tends to resist being moved.
   b. A moving object tends to resist a change in speed.
   c. A moving object tends to resist a change in direction.
   d. all of the above

_____ 2. Which of the following is true of inertia?
   a. It is described in Newton's first law of motion.
   b. It is a property of motion.
   c. It is measured by weight.
   d. all of the above

_____ 3. Two forces act on an object. The magnitude of the net force acting on the object
   a. equals the sum of the magnitudes of the two forces.
   b. equals the difference in the magnitudes of the two forces.
   c. equals the average of the two forces.
   d. cannot be determined from the given information.

_____ 4. An object is at rest on an incline. The component of gravity acting downward and along the incline on the object is described as $F_{g, x} = +8.0$ N. Which axis is pointing up the incline?
   a. $+x$-axis
   b. $+y$-axis
   c. $-x$-axis
   d. $-y$-axis

_____ 5. A force of 5 N and a force of 15 N acting on an object can produce the following net forces *except*
   a. 20 N.   **c.** 10 N.
   b. 15 N.   **d.** 5 N.

_____ 6. Two dogs are tugging on a toy. One dog pulls with a horizontal force of 14 N due north. The other is pulling due east with a horizontal force of 14 N. What is the direction of the net external horizontal force on the toy?
   a. north
   b. northeast
   c. west
   d. northwest

Forces and the Laws of Motion *continued*

_____ **7.** Which of the following statements describe an object in equilibrium?
   I. The object is at rest.
   II. The object is moving at constant velocity.
   III. The net external force on the object is zero.
   a. I and II
   b. I and III
   c. II and III
   d. I, II, and III

_____ **8.** A single force acting on an object is represented in a free-body diagram of the object by the following force vector: ↑. Which of the following force vectors represents a single force acting on the object that will bring the object into equilibrium?
   a. ↑
   b. →
   c. ↓
   d. ←

9. While sitting in a lounge chair with the back inclined at an angle of 10° from the vertical, you tuck a pillow between your head and the chair. Then you lower the chair back to 40° and doze with the pillow in place. How has the net force on the pillow changed?

_____

_____

10. A rigid rod holds a restaurant sign horizontally from the side of a building. The force due to gravity acting on the sign is 165 N with the angle between the sign and the rod being 50.0°. Ignoring the weight of the rod, what is the magnitude of the tension in the rod?

Forces and the Laws of Motion

Section Quiz: Newton's Second and Third Laws

Write the letter of the correct answer in the space provided.

_____ 1. The change in the horizontal velocity of a 12-kg scooter is +0.5 m/s. What is the net horizontal force acting on it?
   a. +24 N
   b. +6.0 N
   c. greater than 0 N
   d. 0 N

_____ 2. A student holds a 6-N block of wood from a spring balance in an express elevator that maintains constant velocity traveling between floors. A spring scale reading of 5.9 N indicates that the elevator is
   a. starting an ascending trip.
   b. ending an ascending trip.
   c. ending a descending trip.
   d. traveling between floors.

_____ 3. During a three-part circus stunt, a clown holds a ball. The clown then tosses the ball upward. After releasing it, the ball is caught a few moments later from above by another clown on a trapeze. Which set of data could represent the normal force exerted by the ground on the first clown during the stunt? The force due to gravity on the clown is 680 N and that on the ball is 20 N.
   a. 700 N, 695 N, 720 N
   b. 700 N, 695 N, 700 N
   c. 700 N, 705 N, 700 N
   d. 700 N, 705 N, 680 N

_____ 4. In which situation is the net force acting on a car zero?
   a. The car increases speed and changes direction.
   b. The car increases speed but does not change direction.
   c. The car maintains its speed but changes direction.
   d. The car maintains both its speed and direction.

_____ 5. A truck and a car uniformly accelerate from rest to a velocity of 3.0 m/s in equal time intervals. The truck is ten times as massive as the car. Which of the following statements is correct?
   a. The acceleration of the truck is 1/10 that of the car.
   b. The truck travels 1/10 the distance of the car.
   c. The force on the truck is 10 times the force on the car.
   d. all of the above

| **Forces and the Laws of Motion** *continued*

_____ **6.** In an action-reaction pair, the
  a. action force is exerted first.
  b. action force and the reaction force are equal in magnitude and act in the same direction.
  c. action force and the reaction force are contact forces only.
  d. action force and the reaction force act on two different objects.

_____ **7.** A batter strikes a baseball with a bat. Identify an action-reaction pair and describe the forces exerted by each.
  a. The batter exerts a force on the bat; the ball exerts a force on the bat.
  b. The batter exerts a force on the bat; the bat exerts a force on the batter.
  c. The bat exerts a force on the batter; the bat exerts a force on the ball.
  d. The ball exerts a force on the bat; the bat exerts a force on the batter.

_____ **8.** In interactions of action-reaction pairs involving Earth and everyday objects, the effect on Earth's motion is often negligible because
  a. field forces do not obey Newton's third law.
  b. Earth has great inertia.
  c. everyday objects cannot exert forces on Earth.
  d. all of the above

9. Explain how action-reaction pairs keep a book sitting on a table in equilibrium.

_____

_____

_____

_____

_____

10. A child tugs on a rope attached to a 0.62-kg toy with a horizontal force of 16.3 N. A puppy pulls the toy in the opposite direction with a force 15.8 N. What is the acceleration of the toy?

Assessment

Forces and the Laws of Motion

Section Quiz: Everyday Forces

Write the letter of the correct answer in the space provided.

_____ **1.** The acceleration due to gravity
  a. is caused by a field force.
  b. has a magnitude of 9.81 m/s^2 at Earth's surface.
  c. is represented by the symbol, a_g.
  d. all of the above

_____ **2.** The normal force
  a. equals F_g in magnitude.
  b. points vertically upward.
  c. is a scalar quantity.
  d. acts perpendicular to a surface.

_____ **3.** For an object in contact with a given surface, the kinetic friction acting on the object
  a. usually is more than static friction.
  b. is independent of the normal force.
  c. depends on the composition and qualities of the surfaces in contact.
  d. acts perpendicular to the surface.

_____ **4.** The coefficient of kinetic friction
  a. is a type contact force.
  b. is measured in newtons.
  c. depends on the normal force.
  d. depends on the composition and qualities of the surfaces in contact.

_____ **5.** A lubricant decreases kinetic friction between surfaces. The most likely reason it reduces kinetic friction is because the lubricant
  a. reduces the normal force.
  b. changes the composition and qualities of the surfaces in contact.
  c. increases static friction.
  d. produces greater adhesion between the two surfaces.

_____ **6.** A plank is inclined at angle θ, north of west. The direction of the normal force exerted by that plank on an object would be oriented at angle θ,
  a. north of west.
  b. east of north.
  c. south of east.
  d. west of south.

_____ **7.** Which of the following statements are true about $F_{s,\,max}$ in the equation

$$\mu_s = \frac{F_{s,\,max}}{F_n} \ ?$$

I. $F_{s,\,max}$ is exerted perpendicular to the surfaces in contact.
II. $F_{s,\,max}$ represents the maximum value of the force of static friction.
III. On a level surface, the magnitude of $F_{s,\,max}$ equals the magnitude of the applied force that starts the object moving.
  a. I only
  b. II only
  c. I, II, and III
  d. II and III

_____ **8.** All of the following statements about kinetic friction are true *except*
  a. Kinetic friction is calculated by the equation $F_f = \mu_\kappa F_n$.
  b. Kinetic friction is exerted parallel to the surface.
  c. Kinetic friction is exerted opposite the direction of motion.
  d. For an object on an incline, kinetic friction increases as the angle of the incline above the horizontal increases.

9. Only under what conditions is static friction greater than kinetic friction?

_____

_____

_____

10. While deciding where to drive a supporting nail, you are pressing a 18-N picture frame against the wall to keep it from falling. What is the minimum force perpendicular to the wall that you must exert on the frame to keep it stationary? The coefficient of kinetic friction for the two surfaces in contact is 0.42.

Assessment

Work and Energy

Section Quiz: Work

Write the letter of the correct answer in the space provided.

_____ **1.** Which of the following sentences uses *work* in the scientific sense.
  a. Stan goes to work on the bus.
  b. Anne did work on the project for 5 hours.
  c. Joseph found that holding the banner in place was hard work.
  d. An engine does work on a car when the car is moving.

_____ **2.** Work is done on an object
  a. whenever a force acts on the object.
  b. whenever a force is perpendicular to the displacement of the object.
  c. whenever a force causes a displacement of the object.
  d. whenever a net force acts on the object.

_____ **3.** In which of the following cases is *no* work done?
  a. A weightlifter lifts a barbell.
  b. A weightlifter holds a barbell overhead.
  c. A weightlifter slowly lowers a barbell.
  d. A weightlifter drops a barbell and the barbell falls to the ground.

_____ **4.** If the sign of work is negative,
  a. the force is in the same direction as the displacement.
  b. the force is perpendicular to the displacement.
  c. the component of the force that does work is in the direction oppo-
    site the displacement.
  d. the component of the force that does work is perpendicular to the
    displacement.

_____ **5.** A painter lifts a bucket of paint, carries it 5 m horizontally, then sets it
    back down. Which of the following is true?
  a. The force of gravity does negative work when the worker lifts the
    bucket.
  b. The painter does positive work on the bucket when carrying it hori-
    zontally at constant speed.
  c. The painter does positive work on the bucket when setting it down.
  d. No net work is done on the bucket.

_____ **6.** Which equation is used to calculate the work done on an object by a
    force at an angle, θ, to the displacement?
  a. $W = Fd$
  b. $W = Fd\cos\theta$
  c. $W = Fd\sin\theta$
  d. $W = mg\sin\theta$

Work and Energy *continued*

_____ **7.** A joule is equivalent to a
  a. N.
  b. N•m.
  c. N/m.
  d. kg•m/s^2.

_____ **8.** A parachutist falls at a constant speed for 200 m. Which of the following is true?
  a. The force of gravity is the only force doing work on the parachutist.
  b. Air resistance is the only force doing work on the parachutist.
  c. No forces are doing work on the parachutist.
  d. No net work is done on the parachutist.

9. A construction worker lifts a heavy cinder block 1 m off the ground, holds it in place for 3 s, then sets it back down in the same place. Describe the forces doing work on the block and the net work on the block throughout this action.

_____

_____

_____

_____

_____

_____

10. A child pulls a wagon 3.0 m using a force of 55 N at an angle 35° above horizontal. The force of friction on the wagon is 12 N. Calculate the net work done on the wagon.

Work and Energy

Section Quiz: Energy

Write the letter of the correct answer in the space provided.

_____ **1.** Energy that is due to the motion of an object is
  a. kinetic energy.
  b. potential energy.
  c. gravitational potential energy.
  d. elastic potential energy.

_____ **2.** Energy stored in the gravitational field of interacting bodies is
  a. kinetic energy.
  b. nonmechanical energy.
  c. gravitational potential energy.
  d. elastic potential energy.

_____ **3.** Energy associated with a compressed or stretched object is
  a. kinetic energy.
  b. potential energy.
  c. gravitational potential energy.
  d. elastic potential energy.

_____ **4.** How does the kinetic energy of an object change if the object's speed
  doubles?
  a. The kinetic energy decreases to half its original value.
  b. The kinetic energy doubles.
  c. The kinetic energy increases by a factor of 4.
  d. The kinetic energy does not change.

_____ **5.** The work-kinetic energy theorem states that
  a. the net work done on an object equals the kinetic energy of the
    object.
  b. the net work done on an object equals the change in the kinetic
    energy of the object.
  c. the change in the net work done on an object equals the kinetic
    energy of the object.
  d. the change in the net work done on an object equals the change in
    the kinetic energy of the object.

| Work and Energy *continued*

_____ **6.** Friction does −400 J of net work on a moving car. How does this
affect the kinetic energy of the car?
  a. The kinetic energy increases by 400 J.
  b. The kinetic energy decreases by 400 J.
  c. The kinetic energy decreases by 160 kJ.
  d. The kinetic energy does not change.

_____ **7.** Which of the following does *not* affect gravitational potential energy?
  a. an object's mass
  b. an object's height relative to a zero level
  c. the free-fall acceleration
  d. an object's speed

_____ **8.** How does the elastic potential energy in a mass-spring system change
if the displacement of the mass is doubled?
  a. The elastic potential energy decreases to half its original value.
  b. The elastic potential energy doubles.
  c. The elastic potential energy increases or decreases by a factor of 4.
  d. The elastic potential energy does not change.

9. Which has more kinetic energy, a 4.0 kg bowling ball moving at 1.0 m/s or a
1.0 kg bocce ball moving at 4.0 m/s? Explain your answer.

_____

_____

_____

_____

_____

10. A 1.0×10^3 kg sports car is initially traveling at 15 m/s. The driver then
applies the brakes for several seconds so that −25 kJ of net work is done on
the car. Calculate the initial and final kinetic energy of the car.

Assessment

Work and Energy

Section Quiz: Conservation of Energy

Write the letter of the correct answer in the space provided.

_____ 1. Which of the following is true of the conservation of energy in a closed system?
   a. Kinetic energy is always conserved.
   b. Potential energy is always conserved.
   c. Mechanical energy is always conserved.
   d. Total energy is always conserved.

_____ 2. The mechanical energy of a system of objects is
   a. the sum of kinetic energy and gravitational potential energy.
   b. the sum of kinetic energy and elastic potential energy.
   c. the sum of kinetic energy and all relevant forms of potential energy.
   d. the sum of all forms of energy.

_____ 3. Mechanical energy is *not* conserved when
   a. gravitational potential energy is converted to kinetic energy.
   b. kinetic energy is converted to gravitational potential energy.
   c. kinetic energy is converted to elastic potential energy.
   d. friction is not negligible.

_____ 4. In which of the following situations is mechanical energy most likely to be conserved?
   a. A football flies through the air.
   b. A feather falls from the sky.
   c. A skateboard rolls into the grass.
   d. A hockey player digs his skates into the ice.

_____ 5. If mechanical energy is conserved in a system, the energy at any point in time can be in the form of
   a. kinetic energy.
   b. gravitational potential energy.
   c. elastic potential energy.
   d. all of the above

_____ 6. Which of the following is *not* a form of mechanical energy?
   a. kinetic energy
   b. chemical potential energy
   c. gravitational potential energy
   d. elastic potential energy

| **Work and Energy** *continued*

_____ **7.** Which of the following is evidence that frictional forces are present in
a system?
   a. Interactions in the system cause an increase in temperature.
   b. Interactions in the system produce sound.
   c. Mechanical energy is not conserved.
   d. all of the above

_____ **8.** An egg suspended above the ground has 2.0 J of gravitational potential
energy. The egg is then dropped and falls to the ground. What is the
kinetic energy of the egg just as it reaches the ground?
   a. −2.0 J
   b. 0 J
   c. 2.0 J
   d. 4.0 J

9. A tennis ball is thrown up into the air starting from a height of 1.5 m. The ball
reaches a peak height, then falls down to the ground. Assuming air resistance
is negligible, describe the energy transfers that take place during the flight of
the ball. Is mechanical energy conserved in this situation?

_____

_____

_____

_____

_____

10. The tennis ball in question 9 above has a mass of 5.7×10^{-2} kg and has an
initial speed of 2.0 m/s. Calculate the speed of the ball when it hits the ground.
Ignore air resistance.

Work and Energy

Section Quiz: Power

Write the letter of the correct answer in the space provided.

_____ **1.** Which of the following refers to the rate at which energy is transferred?
   a. work
   b. kinetic energy
   c. mechanical energy
   d. power

_____ **2.** Which of the following refers to the rate at which work is done?
   a. energy
   b. kinetic energy
   c. mechanical energy
   d. power

_____ **3.** Which of the following is *not* a valid equation for power?
   a. $P = \dfrac{W}{\Delta t}$

   b. $P = \dfrac{Fd}{\Delta t}$

   c. $P = \dfrac{Fv}{\Delta t}$

   d. $P = Fv$

_____ **4.** The SI unit for power is
   a. N•m.
   b. J.
   c. W.
   d. hp.

_____ **5.** How much work can a motor with a power output of 25 W do in 1 s?
   a. $\dfrac{1}{25}$ J
   b. 1 J
   c. 25 J
   d. 25 W

_____ **6.** If a machine increases the distance over which work is done,
   a. the force required to do the work is less.
   b. the force required to do the work is greater.
   c. the force required to do the work is the same.
   d. the amount of work done is increased.

_____ **7.** If a machine decreases the distance over which work is done,
   a. the force the machine applies is less.
   b. the force the machine applies is greater.
   c. the force the machine applies is the same.
   d. the amount of work done is decreased.

_____ **8.** A 100 W light bulb
   a. converts 100 J of kinetic energy to potential energy each second.
   b. converts 100 J of potential energy to kinetic energy each second.
   c. converts 100 J of mechanical energy to nonmechanical energy each second.
   d. converts 100 J of electrical energy to other forms of energy each second.

9. Describe the relationship between energy, time, and power.

_____

_____

_____

_____

10. An engine uses 29 kN of force to power a car at an average speed of 7.5 m/s. What is the average power output of the engine?

Momentum and Collisions

Section Quiz: Momentum and Impulse

Write the letter of the correct answer in the space provided.

_____ **1.** What is the product of an object's mass and its velocity?
 a. kinetic energy
 b. momentum
 c. impulse
 d. inertia

_____ **2.** Which of the following has the greatest momentum?
 a. a 4.0 kg bowling ball moving at 2.0 m/s
 b. a 0.15 kg baseball moving at 10.0 m/s
 c. a 1.6×10^3 kg car moving at 0.5 m/s
 d. a 0.02 kg bullet moving at 950 m/s

_____ **3.** How does the momentum of an object change if the object's velocity doubles?
 a. The momentum doubles.
 b. The momentum increases by a factor of four.
 c. The momentum decreases by a factor of 1/2.
 d. The momentum decreases by a factor of 1/4.

_____ **4.** What are the units of momentum?
 a. N
 b. J
 c. kg•m/s
 d. kg•m/s^2

_____ **5.** Which of the following can determine the magnitude of the change in an object's momentum?
 a. mass and acceleration
 b. force and time interval
 c. force and distance
 d. acceleration and time interval

_____ **6.** Which of the following is true of changes in momentum?
 a. A small force may produce a large change in momentum by acting over a short time interval.
 b. A small force may produce a large change in momentum by acting over a long distance.
 c. A large force may produce a small change in momentum by acting over a short time interval.
 d. A small force may produce a large change in momentum by acting on a very massive object.

| **Momentum and Collisions** *continued*

_____ **7.** If a net force acts on an object, then the object's momentum
   a. will increase.
   b. will decrease.
   c. will either increase or decrease.
   d. may or may not change.

_____ **8.** Which of the following involves a change in momentum?
   a. A bowling ball rolls down the lane at constant speed.
   b. A car coasts down a hill at constant speed.
   c. A sky diver descends with terminal velocity.
   d. A spacecraft travels at constant speed while slowly losing mass.

9. Define impulse, and state the impulse-momentum theorem.

_____

_____

_____

10. A 1.0×10^4 kg spacecraft is traveling through space with a speed of 1200 m/s relative to Earth. A thruster fires for 2.0 min, exerting a continuous force of 25 kN on the spacecraft in a direction opposite the spacecraft's motion. Calculate the initial momentum and the final momentum of the spacecraft.

Name _____ Class _____ Date _____

Momentum and Collisions

Section Quiz: Conservation of Momentum

Write the letter of the correct answer in the space provided.

_____ **1.** A batter hits a baseball back to the pitcher at the same speed as the pitch. Which of the following is true?
   a. The momentum of the ball is the same before and after the batter hits the ball.
   b. The magnitude of the ball's momentum is greater after the batter hits the ball.
   c. The magnitude of the ball's momentum is less after the batter hits the ball.
   d. The magnitude of the ball's momentum is the same before and after the batter hits the ball.

_____ **2.** A small marble collides with a billiard ball that is initially at rest. Which of the following is true?
   a. The momentum of each object increases.
   b. The momentum of each object decreases.
   c. The momentum of the billiard ball increases, and the momentum of the marble decreases.
   d. The total momentum before and after the collision is zero.

_____ **3.** When two ice skaters initially at rest push off one another, their final momenta are
   a. equal in magnitude and direction.
   b. equal in magnitude and opposite in direction.
   c. in the same direction but of different magnitudes
   d. in opposite directions and possibly of different magnitudes.

_____ **4.** When two objects interact in an isolated system,
   a. the momentum of each object is conserved.
   b. the total momentum of the system is zero.
   c. the total momentum is conserved only if the objects move in opposite directions.
   d. the total momentum is always conserved.

_____ **5.** Which of the following expresses the law of conservation of momentum?
   a. The total momentum of an isolated system is zero.
   b. The total momentum of any system always remains constant.
   c. Every object in an isolated system maintains a constant momentum.
   d. The total momentum of an isolated system remains constant regardless of the forces between the objects in the system.

| Momentum and Collisions *continued*

_____ **6.** Conservation of momentum follows from
   a. Newton's first law.
   b. Newton's second law.
   c. Newton's third law.
   d. the law of conservation of energy.

_____ **7.** A billiard ball hits the edge of another billiard ball that is initially at rest. The second ball moves off at an angle. Which of the following is true?
   a. The momentum of the first ball doesn't change.
   b. The momentum of the second ball doesn't change.
   c. The total momentum of the system increases.
   d. The momentum lost by the first ball is gained by the second ball.

_____ **8.** A croquet ball moving at 2.0 m/s strikes another ball of equal mass. The first ball stops moving after the collision. What is the velocity of the second ball after the collision?
   a. −2.0 m/s
   b. 0 m/s
   c. 2.0 m/s
   d. 4.0 m/s

9. Describe the changes in momentum that take place when two billiard balls of equal mass but moving at different speeds collide head-on.

_____

_____

_____

_____

_____

10. A 55 kg boy running at 2.0 m/s jumps onto a 2.0 kg skateboard. Calculate the final velocity of the boy and the skateboard.

Assessment

Momentum and Collisions

Section Quiz: Elastic and Inelastic Collisions

Write the letter of the correct answer in the space provided.

_____ **1.** Two cars collide, lock bumpers, and move together after the collision. What kind of collision is this?
  a. elastic collision
  b. inelastic collision
  c. perfectly elastic collision
  d. perfectly inelastic collision

_____ **2.** A tennis ball is dropped from 1.0 m, bounces off the ground, and rises to 0.85 m. What kind of collision occurred between the ball and the ground?
  a. elastic collision
  b. inelastic collision
  c. perfectly elastic collision
  d. perfectly inelastic collision

_____ **3.** In what kind of collision is kinetic energy always conserved?
  a. elastic collision
  b. inelastic collision
  c. perfectly elastic collision
  d. perfectly inelastic collision

_____ **4.** Most collisions in the everyday world are
  a. elastic collisions.
  b. inelastic collisions.
  c. perfectly elastic collisions.
  d. perfectly inelastic collisions.

_____ **5.** When an inelastic material is in a collision,
  a. the work done to deform the material is equal to the work done to return the material to its original shape.
  b. the work done to deform the material is equal to the work the material does to other objects in the collision.
  c. the work done to deform the material is equal to the increase in the system's total kinetic energy.
  d. some of the work done to deform the material is converted to other forms of energy.

Momentum and Collisions *continued*

_____ **6.** A helium atom collides with another helium atom in an elastic collision. Which of the following is true?

   a. Both momentum and kinetic energy are conserved.

   b. Momentum is conserved but kinetic energy is not conserved.

   c. Kinetic energy is conserved but momentum is not conserved.

   d. Neither momentum nor kinetic energy is conserved.

_____ **7.** Two playground balls collide in an inelastic collision. Which of the following is true?

   a. Both momentum and kinetic energy are conserved.

   b. Momentum is conserved, but kinetic energy is not conserved.

   c. Kinetic energy is conserved, but momentum is not conserved.

   d. Neither momentum nor kinetic energy is conserved.

_____ **8.** Which of the following is *not* evidence that kinetic energy has been lost in a collision?

   a. The collision produces a sound.

   b. At least one of the objects is deformed after the collision.

   c. At least one of the objects increases in temperature as a result of the collision.

   d. One of the objects is at rest after the collision.

9. Explain the difference between inelastic and perfectly inelastic collisions.

_____

_____

_____

_____

_____

10. A 0.16 kg billiard ball moving to the right at 1.2 m/s has a head-on elastic collision with another ball of equal mass moving to the left at 0.85 m/s. The first ball moves to the left at 0.85 m/s after the collision. Find the velocity of the second ball after the collision, and verify your answer by calculating the total kinetic energy before and after the collision.

Circular Motion and Gravitation

Section Quiz: Circular Motion

Write the letter of the correct answer in the space provided.

_____ **1.** Centripetal acceleration must involve a change in
   a. an object's tangential speed.
   b. an object's velocity.
   c. both an object's speed and direction.
   d. the radius of an object's circular motion.

_____ **2.** What is the speed of an object in circular motion called?
   a. circular speed
   b. centripetal speed
   c. tangential speed
   d. inertial speed

_____ **3.** Which of the following is the correct equation for centripetal acceleration?
   a. $a_c = \dfrac{v_t^2}{r}$

   b. $a_c = \dfrac{v_t}{r}$

   c. $a_c = \dfrac{mv_t^2}{r}$

   d. $a_c = \dfrac{(v_{t,f} - v_{t,i})}{\Delta t}$

_____ **4.** What is the centripetal acceleration of a skater moving with a tangential speed of 2.0 m/s in a circular path with radius 2.0 m?
   a. 1.0 m/s²
   b. 2.0 m/s²
   c. 4.0 m/s²
   d. 8.0 m/s²

_____ **5.** What term describes a force that causes an object to move in a circular path?
   a. circular force
   b. centripetal acceleration
   c. centripetal force
   d. centrifugal force

_____ **6.** A centripetal force acts
   a. in the same direction as tangential speed.
   b. in the direction opposite tangential speed.
   c. perpendicular to the plane of circular motion.
   d. perpendicular to tangential speed but in the same plane.

_____ **7.** Centripetal force can be calculated from centripetal acceleration by
   a. dividing by the mass.
   b. multiplying by the mass.
   c. squaring the acceleration and dividing by the radius.
   d. squaring the acceleration, multiplying by the mass, and dividing by the radius.

_____ **8.** Which of the following is due to inertia?
   a. A ball whirled in a circular motion stays in one plane.
   b. A ball whirled in a circular motion experiences centripetal acceleration directed toward the center of motion.
   c. A ball whirled in a circular motion experiences a centripetal force directed toward the center of motion.
   d. A ball whirled in a circular motion will move off in a straight line if the string breaks.

9. Describe the primary force or forces involved when a car executes a turn. Explain why passengers tend to lean or slide toward the outside of the turn.

_____

_____

_____

_____

_____

_____

10. A 1.3×10^3 kg car traveling with a speed of 2.5 m/s executes a turn with a 7.5 m radius of curvature. Calculate the centripetal acceleration of the car and the centripetal force acting on the car.

Circular Motion and Gravitation

Section Quiz: Newton's Law of Universal Gravitation

Write the letter of the correct answer in the space provided.

_____ **1.** What is the centripetal force that holds planets in orbit?
   a. inertia
   b. gravitational force
   c. planetary force
   d. Kepler's force

_____ **2.** The force that Earth exerts on the moon
   a. is greater than the force the moon exerts on Earth.
   b. is less than the force the moon exerts on Earth.
   c. is equal in magnitude to the force the moon exerts on Earth.
   d. causes tides.

_____ **3.** How does the gravitational force between two objects change if the distance between the objects doubles?
   a. The force decreases by a factor of 4.
   b. The force decreases by a factor of 2.
   c. The force increases by a factor of 2.
   d. The force increases by a factor of 4.

_____ **4.** What does G stand for?
   a. free-fall acceleration
   b. gravitational field strength
   c. the constant of universal gravitation
   d. gravitational force

_____ **5.** Which of the following is an expression of gravitational field strength?
   a. $G\dfrac{m_1m_2}{r}$
   c. $G\dfrac{m_E}{r}$
   b. $G\dfrac{m_1m_2}{r^2}$
   d. $G\dfrac{m_E}{r^2}$

_____ **6.** Tides are caused by
   a. differences in the gravitational force of the moon at different points on Earth.
   b. differences in Earth's gravitational field strength at different points on Earth's surface.
   c. differences in the gravitational force of the sun at different points on Earth.
   d. fluctuations in the gravitational attraction between Earth and the sun.

_____ **7.** When a person holds a ball above Earth's surface, the system contains gravitational potential energy. Where is this potential energy stored?
   a. in the ball
   b. inside Earth
   c. in the person holding the ball
   d. in the gravitational field between Earth and the ball

_____ **8.** Evidence confirms that gravitational mass
   a. depends on gravitational field strength.
   b. varies with location.
   c. depends on free-fall acceleration.
   d. equals inertial mass.

9. Explain why an astronaut weighs less on the moon than on Earth.

_____

_____

_____

_____

10. The moon has a mass of 7.35×10^{22} kg and a radius of 1.74×10^{6} m. What is the gravitational force between the moon and an 85 kg astronaut? $(G = 6.673 \times 10^{-11} \text{ N} \bullet \text{m}^2/\text{kg}^2)$

Assessment

Circular Motion and Gravitation

Section Quiz: Motion in Space
Write the letter of the correct answer in the space provided.

_____ 1. According to Copernicus, how do planets move?
   a. Planets move on small circles called epicycles while simultaneously orbiting Earth.
   b. Planets move in circular orbits around Earth.
   c. Planets move in circular orbits around the sun.
   d. Planets move in elliptical orbits with the sun at one focus.

_____ 2. Kepler's laws of planetary motion reconciled
   a. Ptolemaic theory with Copernican theory.
   b. Ptolemaic theory with Copernicus' data.
   c. Copernican theory with Newton's law of universal gravitation.
   d. Copernican theory with Tycho Brahe's data.

_____ 3. Which of the following correctly expresses Kepler's second law?
   a. Planets travel in elliptical orbits with the sun at one focus.
   b. A planet sweeps out equal areas of its orbit in equal time intervals.
   c. A planet's orbital period is proportional to the planet's distance from the sun.
   d. A planet's orbital period is independent of the planet's mass.

_____ 4. Which of the following correctly expresses Kepler's third law?
   a. $T \propto 1/r$
   b. $T^2 \propto r^2$
   c. $T^2 \propto r^3$
   d. $T^3 \propto r^2$

_____ 5. Newton's law of universal gravitation
   a. can be used to derive Kepler's third law of planetary motion.
   b. can be derived from Kepler's laws of planetary motion.
   c. can be used to disprove Kepler's laws of planetary motion.
   d. does not apply to Kepler's laws of planetary motion.

_____ 6. The speed of an object orbiting another object depends on
   a. only the mass of the orbiting object.
   b. only the mass of the object being orbited.
   c. the masses of each object and the distance between them.
   d. the mass of the object being orbited and the distance between the objects.

_____ **7.** How would the period of an object in a circular orbit change if the radius of the orbit doubled?
   a. The period would increase by a factor of 2.
   b. The period would decrease by a factor of 4.
   c. The period would increase by a factor of $2\sqrt{2}$.
   d. The period would decrease by a factor of $2\sqrt{2}$.

_____ **8.** If you were to stand on a bathroom scale in an elevator that is accelerating downward, the bathroom scale would measure
   a. your weight.
   b. your mass.
   c. the force due to gravity between you and Earth.
   d. the normal force between you and the scale.

9. Explain why an astronaut in orbit experiences apparent weightlessness.

_____

_____

_____

_____

_____

10. A satellite with a mass of 2.5×10^3 kg orbits Earth at an altitude of 139 km. Calculate the orbital period and orbital speed of the satellite. ($m_E = 5.97 \times 10^{24}$ kg; $r_E = 6.38 \times 10^6$ m; $G = 6.673 \times 10^{-11}$ N•m^2/kg^2)

Assessment

Circular Motion and Gravitation

Section Quiz: Torque and Simple Machines
Write the letter of the correct answer in the space provided.

_____ 1. What is a measure of the ability of a force to rotate or accelerate an object around an axis?
   a. centripetal force
   b. lever arm
   c. axis of rotation
   d. torque

_____ 2. Which of the following statements is correct?
   a. The closer the force is to the axis of rotation, the less torque is produced.
   b. The closer the force is to the axis of rotation, the easier it is to rotate the object.
   c. The farther the force is from the axis of rotation, the harder it is to rotate the object.
   d. The farther the force is from the axis of rotation, the less torque is produced.

_____ 3. Where should you push on a door to apply the most torque when opening the door?
   a. close to the top of the door
   b. close to the bottom of the door
   c. close to the hinges of the door
   d. far from the hinges of the door

_____ 4. Wrench A is 12 cm long and wrench B is 24 cm long. For a given input force, how does the maximum torque of wrench A compare to the maximum torque of wrench B?
   a. 1/4 as great
   b. 1/2 as great
   c. the same
   d. 2 times as great

_____ 5. Using a machine can allow you to
   a. do less work to perform a given task.
   b. use less force to do a given amount of work.
   c. decrease both the input force and input distance required to do work.
   d. increase both the input force and input distance required to do work.

Circular Motion and Gravitation *continued*

_____ **6.** What kind of simple machine is like two inclined planes placed back-to-back?
   a. a lever
   b. a screw
   c. a wedge
   d. a wheel and axle

_____ **7.** What does mechanical advantage measure?
   a. the ratio of input force to output force
   b. the ratio of output force to input force
   c. the ratio of work input to work output
   d. the ratio of work output to work input

_____ **8.** What does efficiency measure?
   a. the ratio of input force to output force
   b. the ratio of output force to input force
   c. the ratio of work input to work output
   d. the ratio of work output to work input

9. Explain why no real machine can have an efficiency of 100%.

_____

_____

_____

10. An 85 kg man and his 35 kg daughter are sitting on opposite ends of a 3.00 m see-saw. The see-saw is anchored in the center. If the daughter sits 0.20 m from the left end, how far from the right end would the father have to sit for the see-saw to be in balance? ($g = -9.81$ m/s^2)

Assessment

Fluid Mechanics

Section Quiz: Fluids and Buoyant Force
Write the letter of the correct answer in the space provided.

_____ 1. Which of the following is a fluid at room temperature?
    a. oil
    b. wood
    c. lead
    d. aluminum

_____ 2. Which of the following statements is correct?
    a. Liquids have a definite shape.
    b. Gases have a definite volume.
    c. Gases have a definite shape.
    d. Liquids have a definite volume.

_____ 3. What is true about the volume of displaced fluid for an object that is completely submerged?
    a. The volume of displaced fluid is equal to the object's volume.
    b. The volume of displaced fluid is less than the object's volume.
    c. The volume of displaced fluid is greater than the object's volume.
    d. The volume of displaced fluid is not related to the object's volume.

_____ 4. If an object weighing 50.0 N displaces a volume of water with a weight of 10.0 N, what is the buoyant force on the object?
    a. 60.0 N
    b. 50.0 N
    c. 40.0 N
    d. 10.0 N

_____ 5. Which of the following statements is true about the buoyant force on an object that is floating on the surface of a lake?
    a. The buoyant force is greater than the weight of the object.
    b. The buoyant force is equal to the weight of the fluid displaced.
    c. The buoyant force is the same as when the object is completely submerged.
    d. The buoyant force is less than the density of the water.

_____ 6. In which of the following situations will an object sink?
    a. The mass density of the object is less than the mass density of the fluid.
    b. The buoyant force on the object is equal to the weight of the object.
    c. The mass density of the fluid is less than the mass density of the object.
    d. The weight of the fluid displaced equals the weight of the object.

Fluid Mechanics *continued*

_____ **7.** An uncooked egg sinks in fresh water but floats in salt water. Which of the following expressions about the egg's density (ρ_{egg}) with respect to the density of fresh water (ρ_{fw}) and the density of salt water (ρ_{sw}) is correct?

    a. $\rho_{egg} < \rho_{fw} < \rho_{sw}$ **c.** $\rho_{fw} < \rho_{egg} < \rho_{sw}$

    b. $\rho_{sw} < \rho_{egg} < \rho_{fw}$ **d.** $\rho_{fw} < \rho_{sw} < \rho_{egg}$

_____ **8.** Which statement about an object placed in water is correct?

    a. The apparent weight is always less than the weight of the object in air.

    b. The apparent weight is always equal to the weight of the fluid displaced.

    c. The apparent weight is never equal to zero.

    d. The apparent weight is never greater than the buoyant force.

9. Explain why water as ice is not a fluid but water as a liquid or steam is a fluid.

_____

_____

_____

10. Calculate the buoyant force on a cube of metal with an edge of 1.3 cm that is placed in salt water. The density of the metal is 7.86×10^3 kg/m^3, and the density of the salt water is 1.025×10^3 kg/m^3.

Name _____ Class _____ Date _____

Fluid Mechanics

Section Quiz: Fluid Pressure

Write the letter of the correct answer in the space provided.

_____ **1.** According to Pascal's Principle, how is applied pressure transmitted to every point in a fluid and to the walls of the container holding the fluid?
  a. Pressure decreases toward the container walls.
  b. Pressure increases toward the container walls.
  c. Pressure is equal and uniform throughout the fluid.
  d. Pressure depends on the shape of the container.

_____ **2.** An area of 1.0 m^2 and an area of 1.0 cm^2 have the same atmospheric pressure applied to them. Which of the following statements is correct?
  a. The atmospheric force is greater on the 1.0 m^2 area.
  b. The atmospheric force is equal for both areas.
  c. The atmospheric force is less on the 1.0 m^2 area.
  d. Atmospheric force does not depend on pressure or area.

_____ **3.** What force exerts 8.0×10^4 Pa of pressure on an area of 1.0×10^{-2} m^2?
  a. 8.0×10^{-6} N   **c.** 8.0×10^2 N
  b. 8.0×10^{-2} N   **d.** 8.0×10^6 N

_____ **4.** What is pressure that depends on depth, fluid density, and free-fall acceleration called?
  a. total pressure   **c.** absolute pressure
  b. gauge pressure   **d.** atmospheric pressure

_____ **5.** A force of 580 N is applied on a 2.0 m^2 piston of a hydraulic lift. If a crate weighing 2900 N is raised, what is the area of the piston beneath the crate?
  a. 1.0×10^{-2} m^2 **c.** 2.5 m^2
  b. 0.40 m^2 **d.** 1.0×10^1 m^2

_____ **6.** The absolute pressure 20 m beneath the ocean is 3.03×10^5 Pa. Atmospheric pressure above the ocean is 1.01×10^5 Pa. What pressure does the sea water apply?
  a. 4.04×10^5 Pa   **c.** 2.02×10^5 Pa
  b. 3.03×10^5 Pa   **d.** 1.01×10^5 Pa

_____ **7.** The net vertical force due to pressure between two depths within a
fluid equals the weight of the fluid between the two depths. This is
another way of stating which of the following?

    a. Archimedes' principle

    b. Newton's second law

    c. Pascal's principle

    d. the definition of density

_____ **8.** The external pressure crushes a closed vessel when it reaches a
depth of 30.0 m in water ($\rho_w = 1.00$ g/cm^3). Which of the following
statements is true if this same container is immersed in mercury
($\rho_{Hg} = 13.6$ g/cm^3)?

    a. It will be crushed at a greater depth in mercury than in water.

    b. It will be crushed at the same depth in mercury as in water.

    c. It will be crushed at a shallower depth in mercury than in water.

    d. It will not be crushed at any depth.

9. The advantage of a hydraulic lift is that a force applied to a small piston
allows you to lift an object with a weight much greater than the applied force.
However, the smaller piston must be pushed down a farther distance than the
larger piston is raised. Use the concepts of mechanical advantage and Pascal's
principle to explain why this is.

_____

_____

_____

_____

_____

10. A hydraulic lift consists of two cylindrical pistons, one with a radius of 1.5 m
and the other with a radius of 8.0 cm. What force must be applied to the
smaller piston if a crate with a mass of 1.5×10^3 kg is to be raised on the
larger piston?

Assessment

Fluid Mechanics

Section Quiz: Fluids in Motion

Write the letter of the correct answer in the space provided.

_____ **1.** Which of the following statements does *not* describe an ideal fluid?
   a. An ideal fluid is incompressible.
   b. An ideal fluid has internal friction (viscosity).
   c. The flow of an ideal fluid is always steady.
   d. The flow of an ideal fluid is always laminar.

_____ **2.** Which type of flow best describes a river moving through rocky rapids?
   a. steady flow
   b. laminar flow
   c. turbulent flow
   d. viscous flow

_____ **3.** What are the irregular motions in a flowing fluid called?
   a. eddy currents
   b. turbulent currents
   c. laminar flows
   d. nonlinear flows

_____ **4.** Which of the following statements about an ideal fluid moving through a pipe with changing diameter provides the basis for the continuity equation?
   a. The density of the fluid remains constant throughout the pipe.
   b. The mass of the fluid remains constant throughout the pipe.
   c. The volume of the fluid remains constant throughout the pipe.
   d. The speed of the fluid remains constant throughout the pipe.

_____ **5.** A fluid has a flow rate of 5.0 m^3/s as it travels at a speed of 7.5 m/s through a pipe. What is the cross-sectional area of the pipe?
   a. 3.8 m^2
   b. 1.5 m^2
   c. 0.67 m^2
   d. 0.27 m^2

_____ 6. A fluid flows through a pipe whose cross-sectional area changes from 2.00 m^2 to 0.50 m^2. If the fluid's speed in the wide part of the pipe is 3.5 m/s, what is its speed when it moves through the narrow part of the pipe?
    a. 0.071 m/s
    b. 0.88 m/s
    c. 1.1 m/s
    d. 14 m/s

_____ 7. Which of the following is a correct statement of Bernoulli's principle?
    a. The density of a fluid increases as the fluid's velocity increases.
    b. The density of a fluid decreases as the fluid's velocity increases.
    c. The pressure in a fluid increases as the fluid's velocity increases.
    d. The pressure in a fluid decreases as the fluid's velocity increases.

_____ 8. What happens when a breeze blows between two foam plastic balls that are hung by strings 5 cm apart?
    a. The balls move toward each other.
    b. The balls move away from each other.
    c. The balls do not move.
    d. The balls move upward.

9. Use Bernoulli's principle to explain why pressure in a fluid decreases within a pipe as its radius decreases.

_____

_____

_____

_____

10. Use Bernoulli's principle to explain how an airplane achieves lift by moving forward at high speed.

_____

_____

_____

_____

_____

Assessment

Heat

Section Quiz: Temperature and Thermal Equilibrium

Write the letter of the correct answer in the space provided.

_____ 1. Which of the following is proportional to the average kinetic energy of particles in matter?
   a. heat
   b. temperature
   c. thermal equilibrium
   d. internal energy

_____ 2. What is the energy due to both the random motions of a substance's particles and the potential energy due to the bonds between those particles called?
   a. vibrational energy
   b. rotational energy
   c. translational energy
   d. internal energy

_____ 3. What is the type of kinetic energy associated with a molecule spinning about its center of mass called?
   a. vibrational energy
   b. rotational energy
   c. translational energy
   d. internal energy

_____ 4. Which of the following statements best describes a state of thermal equilibrium between two systems?
   a. Both systems have the same mass.
   b. Both systems have the save volume.
   c. Both systems have the same temperature.
   d. Both systems contain the same amount of internal energy.

_____ 5. Which of the following statements correctly describes what occurs to a substance that undergoes thermal expansion?
   a. As the temperature increases, the volume of the substance increases.
   b. As the temperature increases, the volume of the substance decreases.
   c. As the temperature increases, the density of the substance increases.
   d. As the temperature increases, the mass of the substance decreases.

Heat *continued*

_____ **6.** The temperature of the air is measured as 235 K. What is this temperature equal to in degrees Celsius?

   a. 508°C

   b. 203°C

   c. −38°C

   d. −68°C

_____ **7.** How are the Celsius and Kelvin temperature scales similar?

   a. Both scales are based on the freezing and boiling points of water.

   b. Both scales are based on absolute zero.

   c. Neither scale has negative temperature values.

   d. The difference of one degree is the same for both scales.

_____ **8.** Which temperature scale is used widely in science, and is applied to non-scientific uses throughout most of the world?

   a. Celsius

   b. Rankine

   c. Fahrenheit

   d. Kelvin

9. Explain how the kinetic energy of molecules in water accounts for its temperature.

_____

_____

_____

_____

_____

10. The temperature on a warm day is 309.7 K. Calculate the equivalent to this temperature on the Fahrenheit temperature scale.

Name _____ Class _____ Date _____

Heat

Section Quiz: Defining Heat

Write the letter of the correct answer in the space provided.

_____ 1. What term is defined as the energy transferred between objects with different temperatures?
   a. internal energy
   b. work
   c. heat
   d. thermal equilibrium

_____ 2. What must be true if energy is to be transferred as heat between two bodies in physical contact?
   a. The two bodies must have different volumes.
   b. The two bodies must be at different temperatures.
   c. The two bodies must have different masses.
   d. The two bodies must be in thermal equilibrium.

_____ 3. What occurs to the particles in a substance at low temperature when energy is transferred to the substance as heat?
   a. The average kinetic energy of the particles increases.
   b. The average kinetic energy of the particles remains constant.
   c. The average kinetic energy of the particles decreases.
   d. The average kinetic energy of the particles varies unpredictably.

_____ 4. If energy is transferred spontaneously as heat from a substance with a temperature of T_1 to a substance with a temperature of T_2, which of the following statements must be true?
   a. $T_1 < T_2$
   b. $T_1 = T_2$
   c. $T_1 > T_2$
   d. More information is needed.

_____ 5. A container with a temperature of $37°C$ is submerged in a bucket of water with a temperature of $15°C$. An identical container with a temperature of T_1 is submerged in an identical bucket of water with a temperature of T_2. If the amounts of energy transferred as heat between the containers and the water are the same in both cases, which of the following statements is true?
   a. $T_1 = 5°C; T_2 = 37°C$
   b. $T_1 = 10°C; T_2 = 47°C$
   c. $T_1 = 12°C; T_2 = 32°C$
   d. $T_1 = 7°C; T_2 = 29°C$

_____ **6.** What is the process by which energy is transferred by the motion of
cold and hot matter?
   a. thermal conduction
   b. thermal insulation
   c. convection
   d. radiation

_____ **7.** Which of the following is *not* a good thermal insulator?
   a. ceramic
   b. iron
   c. fiberglass
   d. cork

_____ **8.** An arrow strikes a target, causing an internal energy increase of 18 J.
If the arrow is fired so that it enters the target at the same height
above the ground as when it was fired, what is the arrow's initial
kinetic energy?
   a. 0 J
   b. 9 J
   c. 18 J
   d. 27 J

9. Describe how energy is transferred between two objects at different
temperatures, and how they reach thermal equilibrium.

_____

_____

_____

_____

_____

_____

10. A stone with a mass of 0.450 kg is dropped from the edge of a cliff. When the
stone strikes the ground, the internal energy of the stone and ground increase
by 1770 J. If the stone is initially at rest when it is dropped, how high above
the ground is the cliff? ($g = 9.81$ m/s^2)

Assessment

Heat

Section Quiz: Changes in Temperature and Phase

Write the letter of the correct answer in the space provided.

_____ **1.** What is the quantity of energy needed to raise the temperature of a unit mass of a substance by 1°C called?
  a. latent heat
  b. specific heat capacity
  c. internal energy
  d. thermal energy

_____ **2.** Which property of a substance is *not* needed to determine the amount of energy transferred as heat to or from the substance?
  a. temperature change
  b. specific heat capacity
  c. volume
  d. mass

_____ **3.** The specific heat capacity of a substance is determined using a calorimeter containing water. Besides the substance's mass and the change in temperature of the test substance, what other quantities must be measured in calorimetry?
  a. the mass, specific heat capacity, and temperature change of the water
  b. the volume, specific heat capacity, and temperature change of the water
  c. the density, specific heat capacity, and temperature change of the water
  d. the mass, thermal conductivity, and temperature change of the water

_____ **4.** A metal bolt in a calorimeter gives up 3.6×10^3 J of energy as heat to the surrounding water. The bolt has a mass of 0.25 kg and a specific heat capacity of 360 J/kg•°C. What is the change in the bolt's temperature?
  a. 0.40°C
  b. 2.5°C
  c. 4.0°C
  d. 4.0×10^1 °C

| Heat *continued*

_____ **5.** What is the energy transferred to or from a unit mass of a substance during a phase change called?
  a. latent heat
  b. specific heat capacity
  c. internal energy
  d. thermal energy

_____ **6.** During a phase change, which of the following properties does *not* change?
  a. internal energy
  b. physical state
  c. temperature
  d. volume

_____ **7.** In a heating curve, what does a line with a positive slope indicate?
  a. the change in the substance's state with added or removed energy
  b. the increase in the substance's temperature with added energy
  c. the decrease in the substance's temperature with added energy
  d. the change in the substance's latent heat with added energy

_____ **8.** In a heating curve, what does a line with zero slope indicate?
  a. the change in the substance's state with added or removed energy
  b. the increase in the substance's temperature with added energy
  c. the decrease in the substance's temperature with added energy
  d. the change in the substance's specific heat capacity with added energy

9. Using the concept of specific heat capacity, explain why water remains cool on a hot day whereas the air above it becomes hot.

_____

_____

_____

_____

_____

10. A metal part with a mass of 7.50×10^{-2} kg and a temperature of 93.0°C is placed in a calorimeter containing 0.150 kg of water. If the initial temperature of the water is 25.0°C, and the final temperature of the part and water 29.0°C, what is the specific heat capacity of the part? ($c_{p,w} = 4186$ J/kg•°C)

Name _____ Class _____ Date _____

Thermodynamics

Section Quiz: Relationships Between Heat and Work

Write the letter of the correct answer in the space provided.

_____ 1. Which of the following are ways in which energy can be transferred to or from a substance?
  a. heat and internal energy
  b. work and internal energy
  c. heat and work
  d. heat and kinetic energy

_____ 2. What is a set of particles or interacting components that is considered a distinct physical entity called?
  a. an engine
  b. a system
  c. an environment
  d. an ideal gas

_____ 3. How much work is done by 0.020 m^3 of gas if its pressure increases by 2.0×10^5 Pa and the volume remains constant?
  a. 0 J
  b. -4.0×10^3 J
  c. 4.0×10^3 J
  d. 1.0×10^7 J

_____ 4. Pressure equal to 1.5×10^5 Pa is applied to a gas, causing its volume to change by -3.0×10^{-3} m^3. How much work is done?
  a. 0 J
  b. -5.0×10^7 J
  c. -450 J
  d. 450 J

_____ 5. For an isovolumetric process, which of the following statements is correct?
  a. Work, heat, and internal energy all undergo changes.
  b. Work and heat balance each other, so that there is no change in internal energy.
  c. No energy is transferred as heat; internal energy change is due to work.
  d. No work is done; internal energy change is due to heat.

Thermodynamics *continued*

_____ **6.** For an isothermal process, which of the following statements is correct?
  a. Work, heat, and internal energy all undergo changes.
  b. Work and heat balance each other, so that there is no change in internal energy.
  c. No energy is transferred as heat; internal energy change is due to work.
  d. No work is done; internal energy change is due to heat.

_____ **7.** For an adiabatic process, which of the following statements is correct?
  a. Work, heat, and internal energy all undergo changes.
  b. Work and heat balance each other, so that there is no change in internal energy.
  c. No energy is transferred as heat; internal energy change is due to work.
  d. No work is done; internal energy change is due to heat.

_____ **8.** A volume of cool air rapidly descends from the top of a mountain. The air is a poor thermal conductor, but its temperature increases as its volume decreases, because of the increase in atmospheric pressure as it reaches the ground. What thermodynamic process takes place?
  a. adiabatic
  b. isovolumetric
  c. isobaric
  d. isothermal

9. Explain how energy can be transferred as heat and work in an isothermal process and yet the internal energy of the system remains unchanged.

_____

_____

_____

_____

_____

_____

10. Gas within a piston is compressed from a volume of 1.55×10^{-2} m^3 to a volume of 9.5×10^{-3} m^3. What quantity of work is done if the compression occurs under a constant pressure of 3.0×10^5 Pa?

Assessment

Thermodynamics

Section Quiz: The First Law of Thermodynamics

Write the letter of the correct answer in the space provided.

_____ **1.** Which concept does the first law of thermodynamics describe?
    a. conservation of mass
    b. conservation of energy
    c. work-heat equivalence
    d. conservation of momentum

_____ **2.** What occurs when $Q = 0$, so that the first law of thermodynamics takes the form of $\Delta U = -W$?
    a. an isovolumetric process
    b. an isothermal process
    c. an adiabatic process
    d. an isolated system

_____ **3.** What occurs when $W = 0$, so that the first law of thermodynamics takes the form of $\Delta U = Q$?
    a. an isovolumetric process
    b. an isothermal process
    c. an adiabatic process
    d. an isolated system

_____ **4.** What process occurs when $Q = W = 0$, so that the first law of thermodynamics takes the form of $\Delta U = 0$?
    a. an isovolumetric process
    b. an isothermal process
    c. an adiabatic process
    d. an isolated system

_____ **5.** What occurs when $\Delta U = 0$, so that the first law of thermodynamics takes the form of $Q = W$?
    a. an isovolumetric process
    b. an isothermal process
    c. an adiabatic process
    d. an isolated system

_____ **6.** A steam engine takes in 2750 J of energy as heat, gives up 1550 J of energy as heat to its surroundings, and does 850 J of work. What is the change in the internal energy of the engine?
    a. 5150 J
    b. 3450 J
    c. 2050 J
    d. 350 J

Thermodynamics *continued*

_____ **7.** Which of the following statements about ideal cyclic processes is correct?
  a. The energy added as heat is converted entirely to work.
  b. The net work is greater than the net transfer of energy as heat.
  c. The net work done equals the net transfer of energy as heat.
  d. The net work is less than the net transfer of energy as heat.

_____ **8.** A refrigerator removes a quantity of energy as heat from inside the refrigerator and transfers this energy to the air outside the refrigerator. Which statement correctly describes how this is done?
  a. Work is done on the system, so that energy can be transferred as heat from a low temperature to a high temperature.
  b. Work is done by the system, so that energy can be transferred as heat from a low temperature to a high temperature.
  c. Energy is transferred as heat spontaneously from the high temperature inside the refrigerator to the low temperature outside.
  d. Energy is transferred as heat spontaneously from the low temperature inside the refrigerator to the high temperature outside.

9. Explain how a cyclic process resembles and differs from an isothermal thermodynamic process.

_____

_____

_____

_____

_____

10. An engine takes in 6.60×10^5 J of energy as heat and gives up 4.82×10^5 J of energy as heat to the surroundings. Because it is not an ideal engine, its internal energy increases by 4.2×10^4 J. How much work does the engine do?

Assessment

Thermodynamics

Section Quiz: The Second Law of Thermodynamics

Write the letter of the correct answer in the space provided.

_____ 1. Which of the following statements results from the second law of thermodynamics?
  a. Energy added to a system as heat is entirely converted to work.
  b. Energy added to a system as heat cannot be converted entirely to work.
  c. Energy added to a system as heat equals the increase in entropy.
  d. Energy added to a system as heat equals the heat given up to the surroundings.

_____ 2. Which of the following situations would cause an ideal heat engine to do more work and still perform a cyclic process?
  a. increase the heat from the high-temperature reservoir
  b. decrease the heat from the high-temperature reservoir
  c. increase the heat removed to the low-temperature reservoir
  d. prevent any heat from being removed to the low-temperature reservoir

_____ 3. A power plant transfers 82 percent of the energy produced from burning fuel to convert water to steam. Of the energy transferred as heat by the steam, 36 percent is converted doing work through a spinning turbine. Which equation best describes the overall efficiency of the heat-to-work conversion in the plant?
  a. *eff* > 82 percent
  b. *eff* = 82 percent
  c. *eff* = 36 percent
  d. *eff* < 36 percent

_____ 4. If two-thirds of the energy added to an engine as heat is removed to the engine's surroundings as heat, what is the efficiency of the engine?
  a. 1
  b. 2/3
  c. 1/2
  d. 1/3

_____ 5. If the energy added to an engine as heat is 2.5×10^4 J and the net work done by the engine is 7.0×10^3 J, what is the engine's efficiency?
  a. 0.72
  b. 0.39
  c. 0.28
  d. 0.22

Thermodynamics *continued*

_____ **6.** Work was done to freeze 1 kg of water. Which statement best
describes what happened to this system?
    a. Its entropy increased.
    c. Its entropy remained constant.
    b. Its entropy decreased.
    d. Freezing occurred spontaneously.

_____ **7.** What must be true of the combined entropy of a system and
its environment?
    a. It increases.
    b. It remains constant.
    c. It decreases.
    d. It decreases at first, then increases.

_____ **8.** If the entropy of a system decreases during a thermodynamic process,
what must be true about the change in the entropy of the environment?
    a. It increases.
    b. It remains constant.
    c. It decreases.
    d. It decreases at first, then increases.

9. Use the second law of thermodynamics to explain why a heat engine cannot
be 100 percent efficient.

_____

_____

_____

_____

_____

10. An engine takes in 7.6×10^5 J of energy as heat and gives up 5.7×10^5 J of
energy as heat to the surroundings. What is the efficiency of the engine?

Vibrations and Waves

Section Quiz: Simple Harmonic Motion

Write the letter of the correct answer in the space provided.

_____ **1.** According to Hooke's law, the force exerted by a spring on an object is proportional to
   a. the mass of the object.
   b. the displacement of the spring.
   c. the length of the spring.
   d. the volume of the object.

_____ **2.** In any system in simple harmonic motion, the restoring force acting on the mass in the system is proportional to
   a. displacement.
   b. the length of a pendulum.
   c. the mass.
   d. frequency.

_____ **3.** The spring constant in a given oscillating mass-spring may be changed by
   a. increasing the mass.
   b. decreasing the mass.
   c. decreasing the initial displacement.
   d. none of the above

_____ **4.** In an oscillating mass-spring system, the velocity of the mass is greatest when the mass is
   a. at the point of maximum displacement.
   b. halfway between the equilibrium point and maximum displacement.
   c. at the point where acceleration is greatest.
   d. at the equilibrium point.

_____ **5.** The period of a pendulum may be decreased by
   a. shortening its length.
   b. increasing the mass of the bob.
   c. moving its equilibrium point.
   d. decreasing the mass of the bob.

_____ **6.** As the swinging bob of a pendulum moves farther from its equilibrium position, the pendulum's _____ increases.
   a. frequency
   b. mass
   c. restoring force
   d. length

_____ **7.** The gravitational potential energy of the bob of a swinging pendulum is at its maximum when the bob is at
   a. maximum displacement.
   b. the equilibrium point.
   c. the center of its swing.
   d. minimum displacement.

_____ **8.** Stretching a spring increases its _____ energy.
   a. mechanical kinetic
   b. gravitational potential
   c. vibrational kinetic
   d. elastic potential

9. What happens to an oscillating mass-spring system as a result of damping?

_____

_____

10. What is the spring constant of a spring that stretches a distance of 8.5 cm when a mass of 370 g is hung on it?

Assessment

Vibrations and Waves

Section Quiz: Measuring Simple Harmonic Motion

Write the letter of the correct answer in the space provided.

_____ 1. In a system in simple harmonic motion, the amplitude depends on
   a. frequency.
   b. wavelength.
   c. the position of the equilibrium point.
   d. maximum displacement from the equilibrium point.

_____ 2. In an oscillating mass-spring system, the distance of the maximum compression of the spring is a measure of
   a. amplitude.
   b. frequency.
   c. period.
   d. equilibrium.

_____ 3. A total of 5 s passes as a child completes one complete swing on a playground swing. The period of the swing is
   a. 1/5 s.
   b. 20 s.
   c. 5 s.
   d. 1 s.

_____ 4. The frequency of a certain pendulum is 0.5 Hz. The period of this pendulum is
   a. 0.2 s.
   b. 0.5 s.
   c. 2 s.
   d. 5 s.

_____ 5. Which of the following factors has the greatest effect on the frequency of a swinging pendulum?
   a. friction
   b. amplitude
   c. mass
   d. length

_____ 6. If a given pendulum is moved from Earth to a location where the gravitational acceleration is greater than Earth's, the frequency of the pendulum's swing will
   a. increase.
   b. decrease.
   c. be unchanged.
   d. vary with the mass of the bob.

_____ **7.** A certain oscillating mass-spring system has a period of 1.2 s with a
1.0 kg mass. What will the period be when a 4.0 kg mass is substituted
for the 1.0 kg mass?
  a. 4.8 s
  b. 2.4 s
  c. 0.6 s
  d. 0.3 s

_____ **8.** You have constructed an oscillating mass-spring system for an experiment. In order to increase the frequency of the system, you could
  a. decrease the initial displacement.
  b. use a greater mass.
  c. use a spring with a higher spring constant.
  d. increase the period of oscillation.

9. A student builds two pendulums, both 40 cm long. One has a 500 g mass and
the other has a 1 kg mass attached. The student starts both pendulums swinging and is surprised to discover that they have identical periods (and frequencies). How could you explain this observation to the student?

_____

_____

_____

10. Calculate the period of a pendulum that is 1.20 m long. What is this pendulum's frequency? Assume that the pendulum is on Earth.

Assessment

Vibrations and Waves

Section Quiz: Properties of Waves

Write the letter of the correct answer in the space provided.

_____ 1. The material through which a mechanical wave travels is
   a. a medium.
   b. empty space.
   c. ether.
   d. air.

_____ 2. When a transverse wave passes through water, water molecules are displaced
   a. permanently in the direction of the wave motion.
   b. permanently in a direction perpendicular to the wave.
   c. temporarily in the direction of the wave motion.
   d. temporarily in a direction perpendicular to the wave.

_____ 3. A wave that is produced by a single motion that does not repeat is a _____ wave.
   a. transverse
   b. continuous
   c. pulse
   d. compression

_____ 4. The distance between two troughs of a transverse wave is the wave's
   a. amplitude.
   b. wavelength.
   c. frequency.
   d. rarefaction.

_____ 5. A _____ wave travels through a medium as a series of compressions and rarefactions.
   a. sine
   b. longitudinal
   c. pulse
   d. transverse

_____ 6. A wave is passing through a uniform medium. As the frequency of this wave increases, its wavelength
   a. depends on amplitude.
   b. decreases.
   c. increases.
   d. does not change.

| Vibrations and Waves *continued*

_____ **7.** Suppose you are dangling your foot in a swimming pool, making ripples (waves) by moving your foot up and down. What could you do to
make the ripples travel faster through the water?
  a. move your foot up and down more frequently
  b. move your foot up and down less frequently
  c. move your foot up and down more strongly
  d. none of the above

_____ **8.** The amplitude of a mechanical wave determines how much
_____ the wave transfers per unit time.
  a. energy
  b. air
  c. wavelength
  d. matter

9. Suppose a wave travels at a velocity, v_A in medium A and a greater velocity,
v_B, in medium B. If wave motion travels from medium A into medium B, how
will the frequency and wavelength change? Explain your reasoning.

_____

_____

_____

10. What is the wavelength of a radio wave from an FM station that broadcasts
at a frequency of 95.5 MHz? The speed of electromagnetic waves in space is
3.00×10^8 m/s.

Name _____ Class _____ Date _____

Vibrations and Waves

Section Quiz: Wave Interactions

Write the letter of the correct answer in the space provided.

_____ 1. When two transverse waves traveling through a medium meet and exactly coincide, the resulting displacement of the medium
   a. is the sum of the displacements of each wave.
   b. is zero.
   c. is always greater than the displacement of either wave alone.
   d. is always destructive.

_____ 2. When two waves having displacements in opposite directions meet, _____ occurs.
   a. complete cancellation
   b. no interference
   c. constructive interference
   d. destructive interference

_____ 3. Two waves meet and interfere constructively. Which one of the following factors increases?
   a. period   **c.** amplitude
   b. frequency   **d.** wavelength

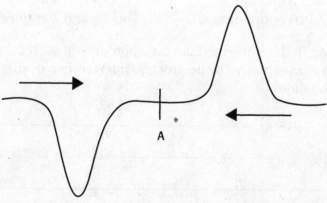

A

_____ 4. The diagram above represents two pulse waves moving toward each other through a medium. The two waves will exactly coincide when they reach point A. At point A, the amplitude of the combined waves will be
   a. twice that of either wave alone.
   b. the same as either wave alone.
   c. half that of either wave alone.
   d. zero.

_____ 5. When a wave on a rope strikes a free boundary, the wave is
   a. reflected and inverted.   **c.** not reflected.
   b. reflected but not inverted.   **d.** absorbed.

Vibrations and Waves *continued*

_____ **6.** A wave pattern on a stretched string appears to be stationary. This wave pattern is an example of a
  a. longitudinal wave.
  b. non-periodic wave.
  c. pulse wave.
  d. standing wave.

_____ **7.** In a standing wave on a vibrating string, a point that appears to remain stationary is a(n)
  a. antinode.
  b. point of constructive interference.
  c. node.
  d. rarefaction.

_____ **8.** The diagram above represents a certain standing wave on a vibrating string. This standing wave has
  a. 4 nodes and 3 antinodes.   **c.** 2 nodes and 3 antinodes.
  b. 3 nodes and 4 antinodes.   **d.** 3 nodes and 2 antinodes.

9. Two waves traveling in the same medium are approaching each other. What determines whether constructive or destructive interference occurs when the waves meet and coincide?

_____

_____

_____

_____

10. A student attaches one end of a rope to a pole and then sends a pulse wave from the opposite end of the rope. Under what conditions will the wave be inverted when it reaches the pole and is reflected?

_____

_____

_____

Assessment

Sound

Section Quiz: Sound Waves

Write the letter of the correct answer in the space provided.

_____ **1.** A sound wave is an example of a(n) _____ wave.
  a. transverse
  b. longitudinal
  c. crest-and-trough
  d. electromagnetic

_____ **2.** A sound wave consists of a series of
  a. compressions and rarefactions.
  b. longitudes and latitudes.
  c. hills and valleys.
  d. perpendicular vibrations

_____ **3.** The human perception of pitch depends on a sound's
  a. velocity.
  b. wavelength.
  c. frequency.
  d. amplitude.

_____ **4.** Sound generally travels faster through solids than through gases
     because the particles (atoms or molecules) of a solid are
     _____ than the particles of a gas.
  a. closer together
  b. heavier
  c. warmer
  d. larger

_____ **5.** Sound waves from a vibrating source travel
  a. in one direction.
  b. in two directions.
  c. in all directions.
  d. back and forth.

_____ **6.** Spherical wave fronts can be treated as parallel lines when
  a. they are very near the source.
  b. the frequency is very high.
  c. the wavelength is very large.
  d. they are a large distance from the source.

Sound *continued*

_____ **7.** When you hear the sound from a vehicle that is moving toward you, the pitch is higher than it would be if the vehicle were stationary. The pitch sounds higher because the
   a. sound waves arrive more frequently.
   b. sound from the approaching vehicle travels faster.
   c. wavelength of the sound waves becomes greater.
   d. amplitude of the sound waves increases.

_____ **8.** Suppose you are on a moving bus as it passes a stationary car in which a stereo is playing music. As the bus goes past the car, you will hear the pitch of the music
   a. become higher.
   b. become lower.
   c. remain constant.
   d. become lower, then higher.

9. Describe how a sound wave is created by an object vibrating in air. Use the words *compression* and *rarefaction* correctly in your explanation.

_____

_____

_____

_____

Wave 1   Wave 2

10. The diagrams above represent sound waves created by two different vibrating objects. Assuming that both waves are traveling through the same medium, which wave is produced by the object vibrating at the higher frequency? Explain your choice.

_____

_____

_____

_____

Assessment

Sound

Section Quiz: Sound Intensity and Resonance
Write the letter of the correct answer in the space provided.

_____ **1.** The measured intensity of sound depends on both the distance from
the source and the _____ of the sound source.
  a. frequency
  b. power
  c. pitch
  d. wavelength

_____ **2.** A sound has an intensity of 1.0×10^{-4} W/m² at a distance of 2.0 m
from the sound source. What is the intensity at a distance of 4.0 m
from the sound source?
  a. 2.0×10^{-4} W/m²
  b. 1.0×10^{-4} W/m²
  c. 5.0×10^{-5} W/m²
  d. 2.5×10^{-5} W/m²

_____ **3.** The perceived loudness of a sound is measured in
  a. watts per square meter.
  b. watts.
  c. decibels.
  d. hertz.

_____ **4.** When the measured intensity of a sound increases from 1.0×10^{-4} to
1.0×10^{-3}, the decibel level increases from 80 dB to
  a. 800 dB.
  b. 100 dB.
  c. 90 dB.
  d. 70 dB.

_____ **5.** The decibel level at the threshold of human hearing is
  a. -1 dB.
  b. 0 dB.
  c. 1 dB.
  d. 10 dB.

_____ **6.** In general, the human ear is most sensitive to sounds having a
frequency of about
  a. 20 Hz.
  b. 250 Hz.
  c. 1000 Hz.
  d. 15 000 Hz.

_____ **7.** When a string on an acoustic guitar is struck, the hollow body of the guitar vibrates at the same frequency as a result of
   a. forced vibration.
   b. compressions.
   c. resonance.
   d. plane waves.

_____ **8.** Resonance occurs when a force causes an object to vibrate at
   a. a large amplitude.
   b. a low pitch.
   c. any frequency.
   d. its natural frequency.

9. Groundskeepers who mow lawns for several hours a day usually wear ear protectors, even though the loudness of the sound from the mower engine is well below the threshold of pain. Why are they wise to wear the protectors?

_____

_____

_____

_____

10. How much power is given off as sound from a gasoline-powered air compressor if the intensity of the sound is 4.6×10^{-3} W/m^2 at a distance of 18 m?

Assessment

Sound

Section Quiz: Harmonics

Write the letter of the correct answer in the space provided.

_____ **1.** A vibrating air column in a cylindrical pipe open on both ends is capable of producing
   a. all frequencies.
   b. only even-numbered harmonics.
   c. only odd-numbered harmonics.
   d. all harmonics.

_____ **2.** The lowest frequency of vibration of a plucked string is the string's _____ frequency.
   a. overtone
   b. timbre
   c. second harmonic
   d. fundamental

_____ **3.** An air column in a pipe vibrates at a fundamental pitch of 680 Hz. If the length of the pipe were doubled, the fundamental pitch would be
   a. 170 Hz.
   b. 340 Hz.
   c. 1360 Hz.
   d. 2720 Hz.

_____ **4.** How many nodes are present in a string vibrating at its fundamental frequency?
   a. 0
   b. 1
   c. 2
   d. 3

_____ **5.** If the first harmonic of a vibrating string has a frequency of 812 Hz, the second and third harmonics will have frequencies of _____, respectively.
   a. 1218 Hz and 1624 Hz
   b. 1218 Hz and 2030 Hz
   c. 1624 Hz and 2436 Hz
   d. 1624 Hz and 3248 Hz

Sound *continued*

_____ **6.** What length of guitar string would vibrate at a fundamental frequency of 1250 Hz if the string is stretched so that the velocity of waves on the string is 488 m/s?
  a. 19.5 cm
  b. 39.0 cm
  c. 1.28 cm
  d. 256 cm

_____ **7.** A term for the quality of sound that gives each different musical instrument a unique sound is
  a. pitch.
  b. overtone.
  c. fundamental.
  d. timbre.

_____ **8.** Two flute players are tuning their instruments. One player sounds a tone with a pitch of 527 Hz and the other player sounds a tone with a pitch of 523 Hz. How many beats per second will the players hear?
  a. none
  b. 2
  c. 4
  d. 525

9. When two notes of slightly different frequencies are sounded, beats may be heard. Explain how these beats occur.

_____

_____

_____

_____

10. The longest common organ pipes are 32 feet (about 10 m) long. What is the fundamental frequency produced by an open-ended organ pipe that is 10.0 m in length? Assume that the pipe is in an environment where the speed of sound is 344 m/s.

Light and Reflection

Section Quiz: Characteristics of Light

Write the letter of the correct answer in the space provided.

_____ **1.** Which of the following is not a component of the electromagnetic spectrum?
    a. light waves
    b. radio waves
    c. microwaves
    d. sound waves

_____ **2.** All of the following statements about electromagnetic waves are true *except* which one?
    a. Electromagnetic waves are distinguished by their different shapes.
    b. Electromagnetic waves are composed of oscillating electric and magnetic fields.
    c. Electromagnetic waves are transverse waves.
    d. Electromagnetic waves move in a direction that is perpendicular to the electric and magnetic fields.

_____ **3.** Given the wave speed equation $c = f\lambda$, what is the relationship between frequency (f) and wavelength (λ)?
    a. direct
    b. exponential
    c. inverse
    d. none of the above

_____ **4.** Which variable in the wave speed equation is constant in a vacuum?
    a. c
    b. f
    c. λ
    d. all of the above

_____ **5.** Which of the following is approximately equal to the currently accepted value for the speed of light?
    a. 3.00×10^6 m/s
    b. 3.00×10^8 m/s
    c. 30.0×10^8 m/s
    d. 3.00×10^9 m/s

_____ **6.** The speed of light is incredibly fast. Physicists consider it to be
    a. finite.
    b. infinite.
    c. immeasurable.
    d. variable.

_____ **7.** The intensity of light depends on the
  a. speed of the light wave and the amount of light energy emitted from a source.
  b. distance from the light source and the type of surface the light strikes.
  c. amount of light energy emitted from a source and the distance from the light source.
  d. size of the surface area which the light strikes and the distance from the light source.

_____ **8.** If the distance from a light source is increased by a factor of 5, the illuminance
  a. also increases by a factor of 5.
  b. decreases by a factor of 25.
  c. increases by a factor of 25.
  d. decreases by a factor of 5.

9. Since each type of electromagnetic wave has its own specific range, explain why the electromagnetic spectrum is continuous rather than divided into distinct sections.

_____

_____

_____

10. What is the frequency of an infrared wave whose wavelength is 650 μm?

Assessment

Light and Reflection

Section Quiz: Flat Mirrors

Write the letter of the correct answer in the space provided.

_____ **1.** Which of the following statements is *not* true about the reflection of light?
  a. All substances absorb at least some incoming light and reflect the rest.
  b. Perfect reflectors are common.
  c. The manner in which light is reflected from a surface depends on the surface's smoothness.
  d. No surface is a perfect reflector.

_____ **2.** What type of light reflection occurs when light is reflected from a rough, textured surface?
  a. diverse
  b. variegated
  c. specular
  d. diffuse

_____ **3.** In how many directions is light reflected from a smooth surface?
  a. one
  b. two
  c. multiple
  d. infinite

_____ **4.** According to the law of reflection, the angle of incidence is _____ the angle of reflection.
  a. less than
  b. equal to
  c. greater than
  d. always twice

_____ **5.** The normal to the surface is drawn _____ to the surface.
  a. parallel
  b. tangential
  c. perpendicular
  d. none of the above

_____ **6.** What is used in order to predict the location of an image?
  a. simple geometry
  b. the law of reflection
  c. ray diagrams
  d. all of the above

Light and Reflection *continued*

_____ **7.** For a flat mirror, object distance (p) is always _____ image
distance (q).
  a. less than
  b. equal to
  c. greater than
  d. half the

_____ **8.** What is the least number of rays needed to locate an image formed by
a flat mirror?
  a. one
  b. two
  c. four
  d. depends on the object

9. What type of image does a flat mirror form? Why?

_____

_____

_____

10. Draw a ray diagram for a 4 cm long pencil located 20 cm in front of a flat
mirror. On the drawing, be sure to label object, image, object distance (p),
and image distance (q).

Light and Reflection

Section Quiz: Curved Mirrors

Write the letter of the correct answer in the space provided.

_____ 1. What type of image is produced by an object that is far from a concave spherical mirror?
  a. smaller and upside down
  b. larger and upright
  c. smaller and upright
  d. larger and upside down

_____ 2. What distinguishes a real image from a virtual image?
  a. Real images are inverted, whereas virtual images are upright.
  b. Real images can be displayed on a surface, whereas virtual images cannot.
  c. Real images can be larger or smaller than the object, whereas virtual images are equal in size to the object.
  d. Real images are possible with any type of mirror, whereas virtual images only occur with flat mirrors.

_____ 3. The mirror equation and ray diagrams are concepts that are valid only for paraxial rays. What is a paraxial ray?
  a. a light ray parallel to the principal axis of the mirror
  b. a light ray perpendicular to the principal axis of the mirror
  c. a light ray very near the principal axis of the mirror
  d. a light ray very far from the principal axis of the mirror

_____ 4. For a spherical mirror, the focal length is equal to _____ the radius of curvature of the mirror.
  a. one-fourth
  b. one-half
  c. twice
  d. the square of

_____ 5. For spherical mirrors, how many reference rays are used to find the image point?
  a. two
  b. three
  c. four
  d. six

Light and Reflection *continued*

_____ **6.** Spherical mirrors suffer from spherical aberration; therefore, the rays
_____ intersect exactly in a single point.
   a. always
   b. often
   c. do not
   d. can

_____ **7.** An object is located in front of a concave spherical mirror between the
center of curvature (C) and the focal point (F). Where is the image
located?
   a. behind the mirror
   b. between the mirror and the focal point
   c. between the focal point and the center of curvature
   d. beyond the center of curvature

_____ **8.** All of the following descriptions about images formed by convex
spherical mirrors are true *except* which one?
   a. They are formed from converging rays.
   b. They are smaller than the objects from which they are formed.
   c. They are always virtual.
   d. Their image distance is always negative.

9. How does a parabolic mirror differ from a spherical mirror? Why is a parabolic mirror often preferred to a spherical mirror?

_____

_____

_____

10. A concave spherical mirror has a focal length of 20.0 cm. Locate the image of
a pen that is placed upright 50.0 cm from the mirror.

Assessment

Light and Reflection

Section Quiz: Color and Polarization
Write the letter of the correct answer in the space provided.

_____ 1. Red, green, and blue are
   a. additive pigments.
   b. subtractive pigments.
   c. additive primary colors.
   d. subtractive primary colors.

_____ 2. How many colors of the visible spectrum can the primary colors form?
   a. all
   b. most
   c. some
   d. the six elementary colors

_____ 3. What color of light is produced when a primary color is combined with its complementary color?
   a. depends on the ratio of the combination
   b. black
   c. a subdued version of the primary color
   d. white

_____ 4. Magenta is the complementary color to green. What two primary colors are used to form magenta?
   a. red and green
   b. red and blue
   c. green and blue
   d. green and cyan

_____ 5. Pigments rely on colors of light that are
   a. added.
   b. subtracted.
   c. reflected.
   d. refracted.

_____ 6. The primary pigments are _____ the primary colors.
   a. complementary to
   b. associated with
   c. the inverse of
   d. unrelated to

Light and Reflection *continued*

_____ **7.** If a light source is emitting electric fields that are oscillating in random directions, the light is said to be
 a. polarized.
 b. diffused.
 c. unpolarized.
 d. scattered.

_____ **8.** The direction in which electric fields are polarized by transmission
 a. is determined by the arrangement of the atoms or molecules in the crystal.
 b. occurs along the transmission axis.
 c. results in linear polarization.
 d. all of the above

9. Why does a proper combination of the three primary pigments produce a black mixture?

_____

_____

_____

10. How can one determine if light is linearly polarized?

_____

_____

_____

Assessment

Refraction

Section Quiz: Refraction

Write the letter of the correct answer in the space provided.

_____ 1. In order for refraction to occur, all of the following conditions must be met *except* which one?
  a. Light rays strike the medium at an angle.
  b. Light crosses the boundary that separates the media.
  c. Light travels from one transparent medium to another.
  d. Light rays strike the medium parallel to the normal.

_____ 2. In which of the following situations is refraction observed?
  a. the virtual image of a plane mirror
  b. polarized light from a car's hood
  c. light entering a pool of water
  d. a real image formed by a concave mirror

_____ 3. When light moves from a material in which its speed is higher into a material in which its speed is lower, the ray is
  a. bent toward the normal.
  b. bent away from the normal.
  c. not bent.
  d. bent toward or away from the normal depending on the light's wavelength.

_____ 4. If light passes from glass into air, the ray will
  a. bend toward the normal.
  b. bend away from the normal.
  c. not bend.
  d. bend toward or away from the normal depending on the light's wavelength.

_____ 5. If an incident light ray strikes a transparent medium parallel to the normal, then refraction
  a. does not occur.
  b. is at its maximum.
  c. depends on the speed of the light in that medium.
  d. none of the above

_____ 6. Which of the following factors change as light waves pass from one transparent medium into another?
  a. f and v
  b. v and λ
  c. f and λ
  d. f, v, and λ

_____ **7.** When a light ray passes from a vacuum into a substance with a large index of refraction, the amount of bending experienced by the ray is
   a. difficult to measure.
   b. small.
   c. large.
   d. insignificant.

_____ **8.** The index of refraction is the ratio of
   a. the wavelength of light versus the frequency of light.
   b. the frequency of light versus the speed of light in a medium.
   c. the wavelength of light versus the speed of light in a vacuum.
   d. the speed of light in a vacuum versus the speed of light in a medium.

9. Use the wave model of light to explain refraction.

_____

_____

_____

10. Find the angle of refraction of a ray of light that enters fused quartz ($n_r = 1.458$) from water ($n_i = 1.333$) at an angle of $35.0°$.

Refraction

Section Quiz: Thin Lenses

Write the letter of the correct answer in the space provided.

_____ **1.** All of the following statements about real and virtual images are true *except* which one?
  a. Real images can be projected onto a screen; virtual images cannot.
  b. Both converging and diverging lenses form real images.
  c. A real image is formed when light rays intersect.
  d. Virtual images form at a point from which light rays appear to come but do not actually come.

_____ **2.** A light ray passing through the center of a thin lens and both of its focal points is
  a. refracted.
  b. not refracted.
  c. not possible.
  d. called the focal ray.

_____ **3.** The size and location of an image formed by a converging lens depends on the
  a. location of the object in relation to the lens.
  b. location of the image in relation to the lens.
  c. location of the observer in relation to the lens.
  d. all of the above

_____ **4.** The image formed by a diverging lens is
  a. always virtual.
  b. always smaller.
  c. always upright.
  d. all of the above

_____ **5.** What relationship does the thin lens equation show?
  a. object size and image size
  b. object stance and image stance
  c. object distance and image distance
  d. object magnification and image magnification

_____ **6.** The resulting magnification (M) of an image formed by a converging lens is 3.0. Therefore, the object is _____ that of the image.
  a. one-ninth
  b. one-third
  c. three times
  d. nine times

| **Refraction** *continued*

_____ **7.** An object is located in front of a converging lens at a distance that is twice the focal length. All of the following descriptions about the image are true *except* which one?
  a. The image is located behind the converging lens at a distance that is twice the focal length.
  b. The image is real.
  c. The image is upright.
  d. The image is the same size as the object.

_____ **8.** A positive sign for magnification (*M*) signifies that the image is
  a. upright and real.
  b. inverted and real.
  c. upright and virtual.
  d. inverted and virtual.

9. How do combination lens systems, such as compound microscopes and refracting telescopes, operate?

_____

_____

_____

10. A candle is located 25.0 cm in front of a converging lens whose focal length is 20.0 cm. Find the image distance.

Assessment

Refraction

Section Quiz: Optical Phenomena

Write the letter of the correct answer in the space provided.

_____ **1.** Which of the following is necessary for light to undergo total internal reflection?
  a. The angle of incidence is less than the critical angle.
  b. Light moves from a medium with a lower index of refraction into a medium with a higher index of refraction.
  c. Light moves from a medium with a higher index of refraction into a medium with a lower index of refraction.
  d. The internal surface of the medium has a reflective surface like a mirror.

_____ **2.** Total internal reflection occurs only if the
  a. index of refraction of the first medium is greater than one.
  b. index of refraction of the second medium is greater than one.
  c. index of refraction of both media are the same.
  d. index of refraction of the first medium is greater than the index of refraction of the second medium.

_____ **3.** If the angle of incidence is greater than the critical angle, the ray is _____ at the boundary.
  a. partially reflected
  b. entirely reflected
  c. not reflected
  d. refracted

_____ **4.** Even when the sun has passed just below the horizon, we can still see the sun for a few more minutes because the sun's light rays are _____ by the atmosphere.
  a. reflected
  b. refracted
  c. inverted
  d. expanded

_____ **5.** A mirage is the result of
  a. warmer air near the surface of Earth.
  b. refraction.
  c. the subconscious addition of a reflecting pool of water.
  d. all of the above

_____ **6.** When white light passes through a prism, what color is bent most?
   a. violet
   b. blue
   c. green
   d. All colors are bent the same amount.

_____ **7.** Various wavelengths of incoming light are bent at different angles
   when they encounter a refracting medium. What is the name of this
   phenomenon?
   a. dispersion
   b. image restructuring
   c. internal reflection
   d. Snell's law

_____ **8.** Which of the following colors of light would bend the most when
   entering a refracting medium?
   a. red
   b. yellow
   c. green
   d. blue

9. What color is observed at the top of a rainbow? Explain.

_____

_____

_____

10. Find the critical angle for an ethyl alcohol-air boundary if the index of refraction of ethyl alcohol is 1.361.

Assessment

Interference and Diffraction

Section Quiz: Interference

Write the letter of the correct answer in the space provided.

_____ **1.** The property by which two monochromatic light waves wavelengths
maintain a constant phase relationship is an example of
   a. coherence.
   b. dispersion.
   c. interference.
   d. refraction.

_____ **2.** If two monochromatic light waves undergo destructive interference,
the amplitude of the resultant wave is
   a. equal to zero.
   b. less than the amplitude of either of the component waves.
   c. greater than the amplitude of either of the component waves.
   d. equal to the sum of the amplitudes of the component waves.

_____ **3.** Which of the following values is the *minimum* phase difference
between two monochromatic light waves that are in phase?
   a. 0°
   b. 45°
   c. 90°
   d. 180°

**Questions 4–7 refer to the following pattern obtained from a double-slit
interference experiment. E is the center of the pattern.**

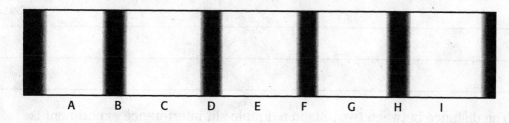

_____ **4.** Fringes _____ form because two light waves that are in phase
interfere.
   a. F and G
   b. G and H
   c. F and H
   d. G and I

_____ **5.** Fringe _____ is a second order maximum.

   a. F   **c.** H

   b. G   **d.** I

_____ **6.** The condition for the second dark fringe in the above pattern is
described by which equation?

   a. $d\sin\theta = \lambda/2$

   b. $d\sin\theta = \lambda$

   c. $d\sin\theta = 3\lambda/2$

   d. $d\sin\theta = 2\lambda$

_____ **7.** Among the changes that occur, which of the following would occur to
the pattern shown in the figure if light of smaller wavelength were
used in the experiment?

   a. E would be displaced to right.

   b. C would be displaced to the right.

   c. G would be displaced to the right.

   d. There would be no change in the pattern.

_____ **8.** A bright fringe appears because of which of the following conditions
of two interfering monochromatic light waves?

   I. A crest of one wave overlaps a crest of another wave.

   II. A crest of one wave overlaps a trough of another wave.

   III. A trough of one wave overlaps a crest of another wave.

   a. I only

   b. II only

   c. III only

   d. I or II

9. Compare and contrast coherent and incoherent waves.

_____

_____

_____

10. The distance between two slits in a double-slit interference experiment is
2.1×10^{-3} mm. A first-order bright fringe is observed at an angle of $17.0°$
from the central maximum. What is the wavelength of the light?

Interference and Diffraction

Section Quiz: Diffraction

Write the letter of the correct answer in the space provided.

_____ 1. Diffraction of waves occurs when a wave encounters which of the
following?
I. obstacle
II. opening
III. sharp edge
IV. another wave of equal wavelength
  a. II only
  b. IV only
  c. I, II, and III
  d. I, II, III, and IV

_____ 2. Which of the following assumptions explains the diffraction of light
waves by a single slit?
  a. Light waves undergo constructive interference only.
  b. Points in a wave front act as point sources of Huygens wavelets.
  c. The width of the slit and the distance to the screen are of the same
  order of magnitude.
  d. all of the above

_____ 3. Which of the following contribute to the constructive interference pattern seen from a compact disc?
I. direction of incoming light
II. orientation of disc
III wavelength of light
  a. I only
  b. II only
  c. I and II
  d. I, II, and III

_____ 4. In an experiment, a variable width slit is illuminated by monochromatic light. As the slit width is narrowed to the same order of magnitude as the wavelength of the light, which of the following will be
observed?
I. The central maximum will become brighter.
II. The central maximum will change color.
III. The secondary maxima will disappear.
  a. I only
  b. II only
  c. II and III
  d. I, II, and III

Interference and Diffraction *continued*

_____ **5.** Which is true about the angle at which a third order maximum is observed if a transmission grating that has a greater number of line per unit length is used.
  a. The angle will increase.
  b. The angle will remain the same.
  c. The angle will decrease.
  d. Third and higher order maxima cannot be observed in patterns formed by transmission gratings.

_____ **6.** What is the function of a spectrometer?
  a. to eliminate light dispersion
  b. to reduce constructive interference
  c. to separate light from a source into its monochromatic components
  d. none of the above

_____ **7.** Which of the following changes would increase the resolution of a telescope?
  a. observing light of greater wavelength, increasing the aperture
  b. observing light of greater wavelength, decreasing the aperture
  c. observing light of lesser wavelength, increasing the aperture
  d. observing light of lesser wavelength, decreasing the aperture

_____ **8.** Instruments that have high resolution are designed to
  a. eliminate coherence of light waves.
  b. enhance dispersion.
  c. maximize diffraction patterns.
  d. reduce interference.

9. If the light passing through a narrow, single slit is coherent, how can a dark fringe appear on a screen?

_____

_____

_____

10. A diffraction grating that has a uniform distance of 4.0×10^{-4} cm between slits is being illuminated by monochromatic light ($\lambda = 6.0 \times 10^2$ nm) shining perpendicular to the grating's surface. A second-order maximum will be observed at what angle from the central maximum?

Interference and Diffraction

Section Quiz: Lasers

Write the letter of the correct answer in the space provided.

_____ **1.** Stimulated emission is the process of
  a. amplifying sound waves.
  b. distinguishing between sources of electromagnetic waves.
  c. producing more light waves that are coherent with an incident light wave.
  d. reducing the energy of mechanical waves.

_____ **2.** A laser includes all of the following *except* a(n)
  a. incandescent filament.
  b. mirror.
  c. partially transparent mirror.
  d. source of energy.

_____ **3.** Characteristics of a beam of light from a laser include all of the following *except*
  a. intense.
  b. monochromatic.
  c. narrow.
  d. non-directional.

_____ **4.** The wavelength of the emitted light from a laser is determined by which of the following?
  a. composition of the active medium
  b. frequency of the amplifying radiation
  c. length of the lasing tube
  d. all of the above

_____ **5.** The intensity of a He-Ne laser is listed as 0.6 W/m^2. Intensity is a measure of which of the following quantities?
  I. energy per area per time
  II. power per area
  III. work per time
  a. I only
  b. I and II
  c. III only
  d. I, II, and III

_____ **6.** Most of the light produced by stimulated emission
  a. emerges as a beam from one end of the laser.
  b. escapes out the sides of the glass tube in a laser.
  c. is reabsorbed by the active medium.
  d. produces greater amounts of coherent light.

_____ **7.** Masers are laser like devices that operate in the _____
region of the electromagnetic spectrum.
   a. microwave
   b. ultraviolet
   c. visible light
   d. X-ray

_____ **8.** Many medical procedures are done with lasers because specific body
tissue
   a. absorbs different wavelength of light.
   b. contains the same amount of red-pigmented blood cells.
   c. is made of similar proteins.
   d. reflects only coherent light.

9. What words are represented by the acronym *laser*?

_____

_____

_____

10. On the second graph below, sketch a wave that is coherent with the wave on
the first graph.

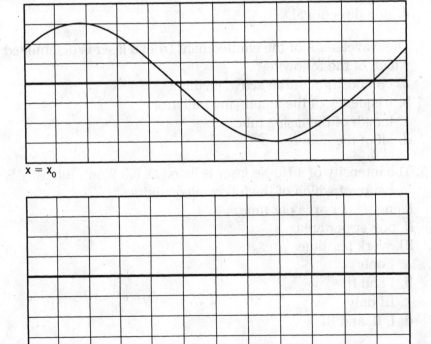

$x = x_0$

$x = x_0$

Electric Forces and Fields

Section Quiz: Electric Charge

Write the letter of the correct answer in the space provided.

_____ **1.** Objects become electrically charged as a result of the transfer of
  a. protons.
  b. electrons.
  c. neutrons.
  d. nuclei.

_____ **2.** The nucleus of an atom has _____ electric charge.
  a. a positive
  b. a negative
  c. a neutral
  d. both positive or negative

_____ **3.** Rubbing a balloon through dry hair gives the balloon a _____
  electric charge.
  a. positive
  b. negative
  c. neutral
  d. balanced

_____ **4.** Robert Millikan's research showed that electric charge is quantized,
  which means that
  a. a proton has a charge equal to that of two electrons.
  b. an electron can have any amount of electric charge.
  c. all charge is an integer multiple of a fundamental charge.
  d. all charge is an integer multiple of the charge of a neutron.

_____ **5.** A student rubs a certain rod with plastic wrap. The rod then repels a
  glass rod that has been rubbed with silk and attracts a rubber rod that
  has been rubbed with fur. This observation shows that the first rod has
  a(n) _____ charge.
  a. positive
  b. negative
  c. neutral
  d. undetermined

_____ **6.** A negatively charged rod is brought near a metal sphere that is *not* grounded. When the rod is taken away, the metal sphere will have
   a. a positive charge.
   b. a negative charge.
   c. an induced charge.
   d. no charge.

_____ **7.** A conductor, such as a copper rod, can be charged by contact with another charged object only if the metal rod is
   a. an insulator.
   b. grounded to Earth.
   c. insulated from Earth.
   d. positively charged.

_____ **8.** Uncharged objects can be attracted by a charged object because the uncharged objects acquire a surface charge by the process of
   a. contact.
   b. polarization.
   c. conduction.
   d. charge transfer.

9. Electrons in conducting materials are loosely held by atoms in contrast to electrons in insulators, which tend to be tightly held. How does this fact explain the difference between electrical conductors and electrical insulators?

_____

_____

_____

10. A grounded conductor may be charged by the process of _____ when a charged object is brought near the conductor.

_____

Assessment

Electric Forces and Fields

Section Quiz: Electric Force

Write the letter of the correct answer in the space provided.

_____ **1.** An electric force F exists between two objects, both having the charge
q. If the charge on one object is doubled to $2q$, the force between the
objects becomes

    a. $\frac{1}{4}F$.
                        c. $2F$.

    b. $\frac{1}{2}F$.
                        d. $4F$.

_____ **2.** An electric force F exists between two objects separated from each
other by distance r. If the distance is decreased to $0.5r$, the force
between the two objects becomes

    a. $\frac{1}{4}F$.
                        c. $2F$.

    b. $\frac{1}{2}F$.
                        d. $4F$.

_____ **3.** Two point charges, initially 2 cm apart, are moved to a distance of 8
cm apart. By what factor does the resulting electric force between
them change?

    a. 2
                        c. $\frac{1}{4}$

    b. 4
                        d. $\frac{1}{16}$

_____ **4.** Two charged objects are near each other. One has a charge of
-2.0×10^{-5} C and the other has a charge of -4.0×10^{-5} C. If the first
charge doubles to -4.0×10^{-5} C, and the second charge doubles to
-8.0×10^{-5}, the force between the charges changes by a factor of

    a. 16
                        c. 2

    b. 4
                        d. No change occurs.

_____ **5.** Electric force and gravitational force are alike in that both forces

    a. depend on charge.
              c. act only when objects are touching.

    b. depend on mass.
                d. are field forces.

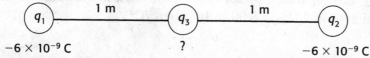

    q_1 1 m q_3 1 m q_2

    -6×10^{-9} C   ? -6×10^{-9} C

_____ **6.** The diagram above represents an array of point charges of the magnitudes shown. Assume that no net force acts on charge q_3. What can
you say about the sign and magnitude of charge q_3?

    a. The charge must be positive and less than the charge on q_1 and q_2.

    b. The charge must be positive and the same as the charge on q_1 and q_2.

    c. The charge must be negative and the same as the charge on q_1 and q_2.

    d. The charge can have any sign and any magnitude.

Electric Forces and Fields *continued*

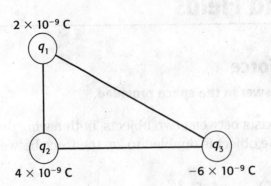

2×10^{-9} C

q_1

q_2

4×10^{-9} C

q_3

-6×10^{-9} C

_____ **7.** The diagram above represents an array of point charges of the magnitudes shown. Which of the following choices best represents the direction of the net electric force acting on charge q_3?

a.

c. q_3

b.

d. q_3

_____ **8.** Which of the following best represents a difference between electric force and gravitational force?
   a. Gravitational force is stronger than electric force.
   b. Electric force can both attract and repel.
   c. Electric force is inversely proportional to the square of the distance between objects.
   d. Gravitational force exists between objects that are not touching.

9. The resultant force acting on a charged object placed near other charged objects is the _____ sum of the of the individual forces acting on the object.

_____

10. Two point charges having charge values of 4.0×10^{-6} C and -8.0×10^{-6} C, respectively, are separated by 2.4×10^{-2} m. What is the value of the mutual force between them? ($k_C = 8.99 \times 10^9$ N•m²/C²)

Assessment

Electric Forces and Fields

Section Quiz: The Electric Field

Write the letter of the correct answer in the space provided.

_____ **1.** Which of the following would be the best to use to determine whether an electric field is present around an object?
   a. a magnetized pin
   b. light from a lamp
   c. a charged table tennis ball
   d. a mass suspended from a spring scale

_____ **2.** Which of the following units is used to state the strength of an electric field?
   a. coulombs per cubic meter
   b. coulombs per newton
   c. newtons per coulomb
   d. meters per coulomb

_____ **3.** The strength of an electric field around a charged object depends on both the magnitude of the charge and
   a. the magnitude of a test charge.
   b. the sign of the object's charge.
   c. the distance from the object.
   d. the volume of space around the object.

_____ **4.** The electric fields of two charges, A and B, are represented by diagrams showing electric field lines. If charge B is greater than charge A, the diagram of charge B will have _____ than the diagram of charge A.
   a. more lines per unit area
   b. longer field lines
   c. straighter field lines
   d. more curved field lines

_____ **5.** The diagram on the right shows the electric field lines around two charges that have been brought near each other. From the diagram you can infer that the charge of the left is _____ than the charge on the right.
   a. negative and smaller in magnitude
   b. negative and larger in magnitude
   c. positive and smaller in magnitude
   d. positive and larger in magnitude

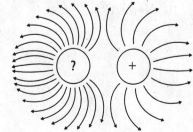

_____ **6.** The diagram on the right represents a cross-section of a charged copper rod. Which of the arrows best represents a field line in the rod's electric field?

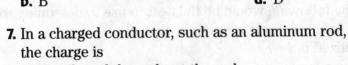

    a. A   **c.** C
    b. B   **d.** D

_____ **7.** In a charged conductor, such as an aluminum rod, the charge is

    a. distributed throughout the rod.
    b. on the surface of the rod.
    c. distributed along a line through the center of the rod.
    d. in the exact center of the rod.

_____ **8.** The diagram on the right represents a spherically shaped conductor with the protrusions shown. If this conductor is given an electrostatic charge, which letter best represents the location of the largest concentration of charge?

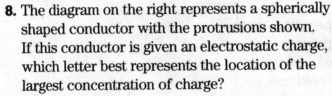

    a. A
    b. B
    c. C
    d. none; The charge is evenly distributed.

9. The drawing on the right is a partial diagram of an electric field showing only a few field lines (without arrowheads). The charge on the right is positive. Can the sign of the charge on the left be determined from the diagram? If so, what is the sign? Explain.

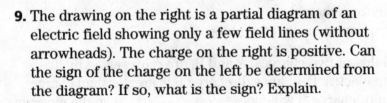

_____

_____

_____

10. An electric field of 2250 N/C is produced by a charge of 4.82×10^{-11} C. For this field strength, what is the distance to the charge? ($k_C = 8.99 \times 10^9$ N•m²/C²).

Assessment

Electrical Energy and Current

Section Quiz: Electric Potential

Write the letter of the correct answer in the space provided.

_____ 1. What is the energy possessed by a charge due to its position in an electric field?
   a. electrical potential energy
   b. electrical kinetic energy
   c. electrical mechanical energy
   d. electrical potential difference

_____ 2. Electric potential
   a. is measured in joules.
   b. depends on the charge at the point where it is measured.
   c. measures energy per unit charge.
   d. is the same as electrical potential energy.

_____ 3. Two positive charges, A and B, are separated by a distance. The electric potential at the position of charge A depends on
   a. the magnitudes of both charges and the distance between them.
   b. the magnitude of charge A and the distance to charge B.
   c. the magnitude of charge B.
   d. the magnitude of charge B and the distance to charge A.

_____ 4. Potential difference is
   a. inversely proportional to change in electrical potential energy.
   b. the measure of the electrical potential energy of a charge.
   c. the ratio of the change in potential energy to the magnitude of a charge.
   d. the ratio of the magnitude of a charge to its change in potential energy.

_____ 5. How does a positive charge move in an electric field in order to gain electrical potential energy?
   a. parallel to the electric field
   b. perpendicular to the electric field
   c. parallel to and in the same direction as the electric field
   d. parallel to and in the opposite direction to the electric field

Electrical Energy and Current *continued*

_____ **6.** A charge moves between two points in a uniform electric field. What information is needed to determine the potential difference between the two points?
  a. the magnitude of the charge, the magnitude of the field, and the displacement in the field
  b. the magnitude of the field and the displacement in the field
  c. the magnitude of the charge and the magnitude of the field
  d. the direction of the field and the displacement in the field

_____ **7.** A battery is a device that maintains a potential difference between two
  a. light bulbs.
  b. terminals.
  c. charges.
  d. chemicals.

_____ **8.** The energy provided by a battery connected to a circuit results from
  a. an electric field inside the battery.
  b. the components of the circuit.
  c. a potential difference.
  d. a chemical reaction.

9. How is the chemical energy in a battery converted to electrical energy?

_____

_____

_____

10. What is the potential difference between a point 0.79 mm from a charge of 7.6 nC and a point at infinity? ($k_C = 8.99 \times 10^9$ N•m^2/C^2)

Electrical Energy and Current

Section Quiz: Capacitance

Write the letter of the correct answer in the space provided.

_____ **1.** What is capacitance?
  a. the amount of charge stored on a conductor
  b. the ability to store energy as separated charges
  c. the ability to store charge on the plates of a capacitor
  d. stored electrical energy

_____ **2.** When a capacitor is connected to a source of potential difference, charges accumulate on the plates of the capacitor. The accumulation of charge stops when
  a. the charge on one plate is equal to the charge on the other plate.
  b. the charge on the capacitor becomes zero.
  c. the potential difference between the plates is the same as the source.
  d. there is a difference in the amount of charge on the two plates.

_____ **3.** Capacitors are devices that store energy. The energy stored in a capacitor is equal to the
  a. charge on the plates of the capacitor.
  b. capacitance of the capacitor.
  c. work required to place the charge on the plates of the capacitor.
  d. difference in potential energy between the plates of the capacitor.

_____ **4.** How does the net charge on a charged capacitor differ from the net charge on an uncharged capacitor?
  a. The net charge on the charged capacitor is greater than on the uncharged capacitor.
  b. There is a net charge on the charged capacitor, but there is no net charge on the uncharged capacitor.
  c. The net charge on both capacitors is the same.
  d. The charged capacitor has a greater positive charge than the uncharged capacitor.

_____ **5.** A material is sometimes placed between the plates of a capacitor to increase the capacitance. What is this material called?
  a. a dielectric
  b. an amplifier
  c. a conductor
  d. a semiconductor

| Electrical Energy and Current *continued*

_____ **6.** Which will increase the capacitance of a capacitor to four times its previous value?
  a. Double the area of both plates.
  b. Increase the area of one plate to four times its previous value.
  c. Reduce the plate spacing by one-half.
  d. Double the area of both plates and reduce the plate spacing by one-half.

_____ **7.** The electrical energy stored in a capacitor is
  a. directly proportional to the voltage across the capacitor.
  b. inversely proportional to the capacitance of the capacitor.
  c. proportional to the square of the voltage across the capacitor.
  d. all of the above

_____ **8.** Which will *not* increase the energy stored in a capacitor by a factor of sixteen?
  a. Increase the capacitance to sixteen times the value.
  b. Increase the potential difference across the capacitor to sixteen times its previous value.
  c. Increase the potential difference across the capacitor to four times its previous value.
  d. Double the potential difference across the capacitor and increase the capacitance by four times.

9. A capacitor is discharged by connecting its terminals with a conducting wire. What happens when the capacitor is discharged?

_____

_____

10. A capacitor has a capacitance of 2200 μF and is charged to a potential difference of 450 V. How much electrical potential energy is stored in the capacitor?

Name _____ Class _____ Date _____

Electrical Energy and Current

Section Quiz: Current and Resistance

Write the letter of the correct answer in the space provided.

_____ **1.** Electric current in a wire is the
   a. number of electric charges moving from one location to another in the wire.
   b. net movement of electrical energy through the wire.
   c. rate at which electric charges move through an area of the wire.
   d. rate at which electrical energy is used to move charges through the wire.

_____ **2.** One ampere of current is the movement of _____ through a given area in one second.
   a. one coulomb of charge
   b. one volt of potential difference
   c. one joule of energy
   d. one electron

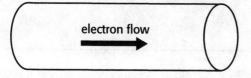

electron flow

_____ **3.** The diagram above represents a wire in which the net flow of electrons is to the right. The direction of the conventional current in the wire
   a. depends on the potential difference.
   b. is to the left.
   c. is the same direction as the electron flow.
   d. cannot be determined.

_____ **4.** The speed at which an electron moves in an electric field in a conductor is the
   a. speed of light.
   b. average collision speed.
   c. drift velocity.
   d. average electron speed.

_____ **5.** When a current moves through a copper conductor, electrons collide with copper atoms. The result of these collisions is
   a. an increase in voltage.
   b. a decrease in resistance.
   c. an increase in temperature.
   d. a decrease in voltage.

_____ **6.** What is the potential difference across a 10 Ω resistor if the current in the resistor is 3.0 A?
  a. 0.3 V
  b. 3 V
  c. 30 V
  d. 15 A

_____ **7.** The current of a non-ohmic material is plotted against the applied potential. How would you describe the resulting graph?
  a. The graph will have a constant positive slope.
  b. The graph will have a variable positive slope.
  c. The graph will have a variable negative slope.
  d. The graph will have a constant negative slope.

_____ **8.** Which factor is *least* likely to affect the resistance of a conductor?
  a. the length of the conductor
  b. the temperature of the conductor
  c. the overall shape of the conductor
  d. the material the conductor is made from

9. When a light switch is turned on, the light comes on almost instantly. Why does the light come on so quickly if the drift velocity of the electrons in the wires is very small?

_____

_____

10. A current of 295 A passes through an automobile starter motor. How long does the starter motor operate if 377 coulombs pass through the starter?

Assessment

Electrical Energy and Current

Section Quiz: Electric Power

Write the letter of the correct answer in the space provided.

_____ 1. A generator converts _____ energy into electrical energy.
   a. chemical
   b. mechanical
   c. kinetic
   d. potential

_____ 2. Electrical current in which there is no net motion of the charge carriers is
   a. direct current.
   b. alternating current.
   c. static current.
   d. static electricity.

_____ 3. In alternating current, the charge carriers move
   b. in all directions.
   a. in only one direction.
   d. repeatedly in one direction then in the opposite direction.
   c. from a lower to a higher electric potential.

_____ 4. Electrical appliances have labels that state the power used by the appliance. What does the power rating listed on the label represent?
   a. the amount of current the appliance uses
   b. the amount of electrical energy converted to heat or light by the appliance
   c. how quickly the appliance heats up
   d. the amount of energy converted each second into other forms of energy

_____ 5. Electric power is the rate at which charge carriers
   a. pass through an area.
   b. do work.
   c. move through a potential difference.
   d. collide with atoms.

_____ 6. The phenomenon called I^2R loss is also known as
   a. joule heating.
   b. resistance.
   c. ohmic energy.
   d. alternating current loss.

_____ **7.** Which statement is correct regarding electric power.
  a. Electric power equals current times resistance.
  b. Electric power equals the amount of electrical energy used by a device.
  c. Electric power equals the current times potential difference.
  d. Electric power equals the amount of electrical energy converted to other forms of energy.

_____ **8.** A certain circuit connected to a 1.5 V battery dissipates 2.0 W of power. If a second battery is added to the circuit, resulting in a potential difference of 3.0 V, what power will be dissipated? Assume that the resistance of the circuit does not change.
  a. 0.50 W
  b. 1.0 W
  c. 2.0 W
  d. 4.0 W

9. Explain how alternating current can deliver energy to a motor without any net flow of electrons through the motor.

_____

_____

10. A microwave oven connected to a 117 V outlet operates at a current of 8.7 A. If the microwave oven is operated exactly 1 hour each day for 28 days, what is the cost of operating this microwave oven if the cost of electric energy is $0.115 per kW•h?

Assessment

Circuits and Circuit Elements

Section Quiz: Schematic Diagrams and Circuits

Write the letter of the correct answer in the space provided.

_____ **1.** A load in a circuit
   a. is a source of potential difference.
   b. dissipates energy.
   c. opens and closes the circuit.
   d. creates electrical energy.

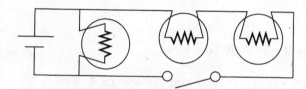

_____ **2.** The symbols in the diagram above indicate
   a. one battery, three resistors, and one closed switch.
   b. one capacitor, three lamps, and one closed switch.
   c. one battery, three lamps, and one open switch.
   d. three resistors, one open switch, and one battery.

_____ **3.** The part of a circuit that converts electrical energy to other forms of
   energy is
   a. a wire.   **c.** the load.
   b. a battery.   **d.** the switch.

_____ **4.** A short circuit is
   a. potentially hazardous.
   b. a circuit in which electrons cannot flow.
   c. a circuit without a load that presents little resistance to electron
   flow.
   d. both a and c

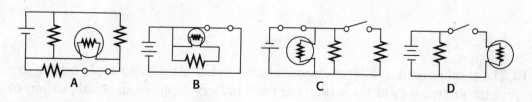

A   B   C   D

_____ **5.** Which of the schematics shown above conducts electricity through the
   lamp?
   a. A and B   **c.** A and C
   b. C and D   **d.** B and D

Circuits and Circuit Elements *continued*

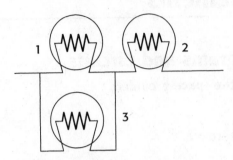

_____ **6.** Three lamps are arranged as shown above. Which single lamp must burn out to cause all of the lamps to go out?
   a. lamp 1
   b. lamp 2
   c. lamp 3
   d. No single lamp will cause all to go out.

_____ **7.** Which is correct regarding the terminal voltage of a battery?
   a. Terminal voltage is always the same as the emf of the battery.
   b. Terminal voltage is always greater than the emf of the battery.
   c. Terminal voltage is always less than the emf of the battery.
   d. The emf of the battery is the potential difference across its terminals.

_____ **8.** A battery is used to supply power to a portable MP3 player. If the terminal voltage across the battery is 4.5 V, what is the potential difference across the MP3 player?
   a. 1.5 V
   b. 4.5 V
   c. 9.0 V
   d. 13.5 V

9. Draw a schematic diagram containing one multiple cell battery, one lamp, one resistor, and one open switch.

10. The total potential difference across the ten batteries that supply power to a portable two-way radio is 12.0 V. If each battery supplies an equal portion of the potential difference, what is the potential difference across each battery, and what is the potential difference across the radio?

Assessment

Circuits and Circuit Elements

Section Quiz: Resistors in Series or in Parallel

Write the letter of the correct answer in the space provided.

_____ **1.** Several resistors are wired in a circuit so that there is a single path for the flow of electric current. What type of circuit is this?
  a. electronic circuit
  b. series circuit
  c. parallel circuit
  d. short circuit

_____ **2.** Five resistors are wired in a series circuit. How does the equivalent resistance of the circuit compare to the resistances of the individual resistors?
  a. The equivalent resistance is greater than any single resistance.
  b. The equivalent resistance is less than any single resistance.
  c. The equivalent resistance is the same as any single resistance.
  d. The equivalent resistance is one-fifth the value of any single resistance.

_____ **3.** Several resistors are wired in series. What is true about this circuit?
  a. The sum of the currents through each of the resistors is equal to the total circuit current.
  b. The total circuit current is the same as the current through any one of the resistors.
  c. The voltage across any resistor is the same as the voltage of the power supply.
  d. The current in any single resistor is determined by its resistance and the voltage of the power supply.

_____ **4.** Two resistors and a battery are wired in a series circuit. One resistor has twice the resistance of the other resistor. What is true about the voltage across the two resistors?
  a. Each resistor has half the battery voltage across it.
  b. One-third of the battery voltage is across the higher value resistor.
  c. One-third of the battery voltage is across the lower value resistor.
  d. Two-thirds of the battery voltage is across the lower value resistor.

_____ **5.** The current through one resistor in a parallel resistor circuit is always
  a. the same as the current in the other resistors in the circuit.
  b. less than the total current in the circuit.
  c. equal to the total current in the circuit.
  d. more than the total current in the circuit.

_____ **6.** What distinguishes a parallel circuit from a series circuit?
  a. The current in a parallel circuit is greater than in a series circuit.
  b. The equivalent resistance of a parallel circuit is less than that of a series circuit.
  c. A parallel circuit always has more than one current path.
  d. The voltage across the resistors in a parallel circuit is greater than it is in a series circuit.

_____ **7.** Four resistors having equal values are wired as a parallel circuit. How does the equivalent resistance of the circuit compare to the resistance of a single resistor?
  a. The equivalent resistance is greater than the resistance of any single resistor.
  b. The equivalent resistance is the same as the resistance of any single resistor.
  c. The equivalent resistance is one-fourth the resistance value of a single resistor.
  d. The equivalent resistance is one-half the resistance value of a single resistor.

_____ **8.** Six resistors are wired in a parallel circuit. What is the voltage across each resistor in the circuit if the first resistor is connected to a 24 V battery?
  a. 4 V
  b. 24 V
  c. 0.25 V
  d. Voltage cannot be determined without the resistance values.

9. A parallel circuit is composed of any number of resistors all of equal value. What simple process can you use to determine the equivalent resistance of such a circuit?

_____

_____

10. Three resistors are wired in parallel with a battery. Two of the resistors have resistances of 38.7 Ω, and 89.5 Ω. The current in the 38.7 Ω resistor is 0.155 A and the total circuit current is 0.250 A. What is the resistance of the third resistor?

Circuits and Circuit Elements

Section Quiz: Complex Resistor Combinations

Write the letter of the correct answer in the space provided.

_____ **1.** You want to determine the current in a complex circuit. Which piece of information will be *least* helpful in making your determination?
  a. the equivalent resistance of the circuit
  b. the number of devices in the circuit
  c. the current in each element in the circuit
  d. the voltage across each element in the circuit

_____ **2.** You have three 100 Ω resistors available. How would you connect these three resistors to produce a 150 Ω equivalent resistance? You must use all of the resistors.
  a. Connect all three resistors in parallel.
  b. Connect all three resistors in series.
  c. Connect one resistor in series with two resistors in parallel.
  d. Connect two resistors in series with the third resistor in parallel to the first two.

_____ **3.** A circuit is constructed as follows: four resistors in parallel connected in series with three resistors in parallel connected in series with two resistors in parallel. All of the resistors have the same value. How does the equivalent resistance of this circuit compare to the resistance of a single resistor?
  a. The equivalent resistance is less than a single resistor.
  b. The equivalent resistance is the same as a single resistor.
  c. The equivalent resistance is greater than a single resistor.
  d. The equivalent resistance cannot be determined without the resistance value of a single resistor.

_____ **4.** Because household devices are connected in parallel in a circuit, _____ as new devices are connected.
  a. the current does not increase
  b. the potential difference remains the same
  c. the resistance of the circuit increases
  d. the power dissipated in the circuit decreases

_____ **5.** In any complex resistance circuit, the current through any resistor in
the circuit is always
  a. less than the total current through the circuit.
  b. equal to the total circuit current.
  c. less than or equal to the total circuit current.
  d. less than or greater than the total circuit current.

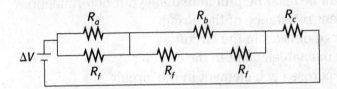

_____ **6.** In the circuit shown above, which resistors, if any, have equal voltages
across them?
  a. R_e and R_f **c.** R_a, R_b, and R_c
  b. R_a and R_d **d.** R_b, R_e, and R_f

_____ **7.** A circuit breaker is designed to limit the amount of _____ in a
circuit.
  a. current   **c.** resistance
  b. voltage   **d.** potential difference

_____ **8.** Which of the following devices is always connected in series with a
household circuit?
  a. air conditioner   **c.** lamp
  b. electric range   **d.** fuse

9. Briefly describe how to find the equivalent resistance of a complex circuit.

_____

_____

_____

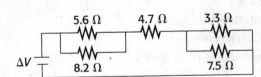

10. Five resistors are wired as shown in the figure above. If the current in the
4.7 Ω resistor is 0.44 A, what is the terminal voltage of the battery?

Assessment

Magnetism

Section Quiz: Magnets and Magnetic Fields

Write the letter of the correct answer in the space provided.

_____ 1. Which of the following statements correctly describes the behavior of magnets?
   a. Like poles attract each other, and unlike poles repel each other.
   b. Like poles repel each other, and unlike poles attract each other.
   c. Both like and unlike poles can attract and repel each other depending on the surrounding materials.
   d. none of the above

_____ 2. What do physicists call large groups of atoms whose net spins are aligned because of strong coupling between neighboring atoms?
   a. magnetic zones
   b. magnetic regions
   c. magnetic sectors
   d. magnetic domains

_____ 3. Which of the following statements about magnetic fields, **B**, is *not* true?
   a. Magnetic fields are vector quantities.
   b. Magnetic fields have both magnitude and direction.
   c. Magnetic field strength increases as the distance from the magnetic source increases.
   d. Magnetic fields are regions in which magnetic forces can be detected.

_____ 4. How is the direction of a magnetic field, **B**, defined at any location?
   a. the direction toward which the south pole of a compass needle points
   b. the direction toward which the north pole of a compass needle points
   c. the direction that is parallel to the imaginary magnetic field lines
   d. the direction that is perpendicular to Earth's magnetic field

_____ 5. What describes Φ_M?
   a. magnetic flux
   b. the number of field lines that cross a certain area
   c. $AB\cos\theta$
   d. all of the above

Magnetism *continued*

_____ **6.** Since more magnetic field lines cross the area that is near the pole of a magnet, what does this indicate about the magnetic field strength in that location?

a. It is stronger.

b. It is weaker.

c. It is entering the magnet.

d. It is leaving the magnet.

_____ **7.** Which of the following statements about magnetic field lines is *not* true?

a. Magnetic field lines form open or closed loops.

b. Magnetic field lines appear to begin at the north pole of a magnet.

c. Magnetic field lines have no beginning or end.

d. Magnetic field lines appear to end at the south pole of a magnet.

_____ **8.** Which of the following statements about the orientation and effects of Earth's magnetic field is *not* true?

a. The geographic North Pole of Earth is near the magnetic south pole.

b. A compass needle that can rotate both perpendicularly and parallel to the surface of Earth, points down at the magnetic south pole.

c. A compass needle always indicates the direction of true north.

d. The geographic South Pole of Earth corresponds to the magnetic north pole.

9. For each of the figures below, indicate whether the magnets will attract or repel one another.

a. | N   S | | N   S |

b. | S   N | | S   N |

c. | S | | S | **d.** | N | | S |
      | N | | N | | S | | N |

10. Draw magnetic field lines around the magnet below. Indicate the relative strength of the magnetic field by drawing more lines where the magnetic field is strongest.

| N   S |

Assessment
Magnetism

Section Quiz: Magnetism from Electricity
Write the letter of the correct answer in the space provided.

_____ **1.** Which of the following terms correctly describes the shape of the magnetic field around a long, straight current-carrying wire?
  a. cylindrical
  b. parallel
  c. perpendicular
  d. elliptical

_____ **2.** In what direction do compass needles deflect in relation to the concentric circles of the magnetic field that is found around a current-carrying wire?
  a. away from the concentric circles
  b. perpendicular to the concentric circles
  c. toward Earth's North Pole irrespective of the concentric circles
  d. tangential to the concentric circles

_____ **3.** The right-hand rule states that the thumb of the right hand is to be placed in what direction when grasping a wire?
  a. in a direction opposite of the current
  b. in the direction of the current
  c. in the direction of Earth's magnetic north pole
  d. none of the above

_____ **4.** According to the right-hand rule, in what direction will the fingers curl?
  a. in the direction of the current
  b. in the direction of the magnetic field, **B**
  c. in the direction of the magnetic field's movement
  d. in the direction of the magnetic field's force

_____ **5.** Which of the following statements about the magnetic field, **B**, around a current-carrying wire is *not* true?
  a. **B** is proportional to the current in the wire.
  b. **B** is inversely proportional to the distance from the wire.
  c. The lines of **B** form concentric circles about the wire.
  d. **B** is independent of the current in the wire.

_____ **6.** The magnetic field of a current loop is most similar to that of a(n)
  a. horseshoe magnet.
  b. circular magnet.
  c. bar magnet.
  d. irregular magnet.

_____ **7.** The strength of a solenoid can be increased by
   a. increasing the current.
   b. increasing the number of wire loops.
   c. inserting an iron rod through the center of the coil.
   d. all of the above

_____ **8.** Which of the following statements about the magnetic field lines of a
   solenoid is *not* true?
   a. The field lines inside a solenoid point in the same direction and are
   nearly parallel.
   b. The field lines inside a solenoid are uniformly spaced and close
   together.
   c. The field lines outside a solenoid are uniformly spaced but further
   apart.
   d. The field lines outside a solenoid do not always point in the same
   direction.

9. What is a solenoid? What is a solenoid called when an iron rod is inserted in
   its center? What advantage is gained by this addition?

   _____

   _____

   _____

   _____

10. Draw magnetic field lines around the figure below and label its north and
   south magnetic poles.

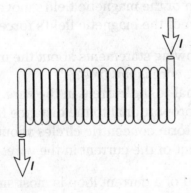

Assessment

Magnetism

Section Quiz: Magnetic Force

Write the letter of the correct answer in the space provided.

_____ **1.** When does the magnetic force on a charge moving through a constant magnetic field reach its maximum value?
   a. When the charge moves parallel to the magnetic field.
   b. When the charge moves at any angle to the magnetic field.
   c. When the charge moves perpendicular to the magnetic field.
   d. When the charge is stationary to the magnetic field.

_____ **2.** Given the following equation, $B = \frac{F_{magnetic}}{qv}$, which of the following statements is true?
   a. The magnetic field, **B**, is directly proportional to q.
   b. The magnetic field, **B**, is directly proportional to v.
   c. The magnetic field, **B**, is directly proportional to $F_{magnetic}$.
   d. The magnetic field, **B**, is directly proportional to $F_{magnetic}$, q, and v.

_____ **3.** If $F_{magnetic} = 3.8 \times 10^{-13}$ N, $q = 1.60 \times 10^{-19}$ C, and $v = 2.4 \times 10^6$ m/s, use the equation, $B = \frac{F_{magnetic}}{qv}$, to find the magnitude of the magnetic field, **B**.
   a. 9.9×10^{11} T
   b. 9.9×10^{-27} T
   c. 5.7×10^{39} T
   d. 0.99 T

_____ **4.** Which of the following statements about the right-hand rule is *not* true?
   a. The fingers indicate the direction of the magnetic field.
   b. The thumb indicates the direction of a particle's movement in the magnetic field.
   c. The direction of the magnetic force is always parallel to the magnetic field.
   d. The direction of the magnetic force exerted on a proton is out of the palm of the hand.

_____ **5.** According to the right-hand rule, what is the direction of the force on an electron?
   a. out of the palm of the hand
   b. out of the back of the hand
   c. in the direction of the thumb
   d. in the direction of the fingers

| **Magnetism** *continued*

_____ **6.** What path does a positively charged particle moving perpendicularly to a uniform magnetic field follow?
   a. erratic
   b. helical
   c. straight
   d. circular

_____ **7.** A current is moving from north to south through a long wire that is lying horizontally on a table. What is the direction of the magnetic force if the magnetic field is directed up and out of the table?
   a. toward the north
   b. toward the east
   c. toward the south
   d. toward the west

_____ **8.** Given $B = 6.0 \times 10^{-4}$ T, $I = 25$ A, and $\ell = 10.0$ m, find $F_{magnetic}$ using the following equation, $F_{magnetic} = BI\ell$?
   a. 1.5×10^{-1} N
   b. 1.5×10^{7} N
   c. 2.4×10^{2} N
   d. 2.4×10^{-4} N

9. Why does a current-carrying wire experience a magnetic force when the wire is placed perpendicular to the magnetic field?

_____

_____

_____

10. Determine the velocity of a proton that is moving perpendicular to a magnetic field whose magnitude is 3.5×10^{-3} T and magnetic force is 8.2×10^{-16} N. Recall that a proton's charge is 1.60×10^{-19} C.

Assessment

Electromagnetic Induction

Section Quiz: Electricity from Magnetism

Write the letter of the correct answer in the space provided.

_____ 1. Under which of the following situations can a current be induced in a circuit?
   a. when an open circuit moves through a magnetic field
   b. when both a closed circuit and magnetic field are moving, but not with respect to each other
   c. when neither the closed circuit nor magnetic field are moving
   d. when a closed circuit moves through a magnetic field

_____ 2. How can the magnitude of the induced emf be increased?
   a. Decrease the length of the wire.
   b. Increase the velocity of the wire as it moves through the magnetic field.
   c. Decrease the strength of the magnetic field.
   d. none of the above

_____ 3. The induced current in a closed loop of wire is zero if
   a. the plane of the loop is perpendicular to the magnetic field.
   b. the plane of the loop is tilted into the magnetic field.
   c. the plane of the loop is parallel to the magnetic field.
   d. the plane of the loop rotates in the magnetic field.

_____ 4. In what way can a current be induced in a circuit?
   a. Move the circuit into or out of a magnetic field.
   b. Vary the intensity and/or direction of the magnetic field.
   c. Rotate the circuit in the magnetic field.
   d. all of the above

_____ 5. Which of the following conditions will decrease the magnitude of an induced current through a circuit loop rotating in a magnetic field?
   a. Reduce the loop's rate of rotation.
   b. Increase the size of the loop.
   c. Increase the magnetic field strength.
   d. Reduce the number of loops that are parallel to the magnetic field.

_____ **6.** When a bar magnet and coil approach each other, the magnetic field produced by the induced current in the coil _____ the approaching magnetic field of the bar magnet.
   a. attracts
   b. repels
   c. inverts
   d. displaces

_____ **7.** How can the polarity or direction of an induced current be reversed?
   a. by reversing the direction of the conductor's motion
   b. by reversing the direction of the magnetic field's motion
   c. either a or b
   d. It cannot be reversed.

_____ **8.** Lenz's law states that
   a. the magnetic field of the induced current is in a direction to produce a field that opposes the change causing it.
   b. the magnetic field of the induced current is in a direction to produce a field that cancels out the change causing it.
   c. the magnetic field of the induced current is in a direction to produce a field that supersedes the change causing it.
   d. none of the above

9. How is it possible to induce a current in a wire that moves through a magnetic field without the use of a battery or an electrical power supply?

_____

_____

_____

_____

10. A 150-turn coil of wire moves perpendicularly through a uniform magnetic field in 1.1 s. If the field increases uniformly from 0.00 T to 0.95 T and each coil of wire has an area of 1.4 m^2, use Faraday's law of magnetic induction to find the magnitude of the induced emf in the coil.

Assessment

Electromagnetic Induction

Section Quiz: Generators, Motors, and Mutual Inductance

Write the letter of the correct answer in the space provided.

_____ **1.** A generator converts
  a. electrical energy into mechanical energy.
  b. mechanical energy into electrical energy.
  c. mechanical energy into steam.
  d. steam into mechanical energy.

_____ **2.** Which of the following statements best describes how a generator
  induces a current?
  a. The magnetic field strength is varied.
  b. A wire loop is moved in and out of the magnetic field.
  c. The orientation of the loop is changed with respect to the
  magnetic field.
  d. The rotation of the loop is reversed periodically.

_____ **3.** The greatest induced emf occurs when segments of a wire loop are
  a. moving parallel to the magnetic field.
  b. moving perpendicularly to the magnetic field.
  c. decelerating in the magnetic field.
  d. accelerating in the magnetic field.

_____ **4.** Maximum emf occurs when the plane of a wire loop is
  a. parallel to the magnetic field.
  b. perpendicular to the magnetic field.
  c. decelerating in the magnetic field.
  d. accelerating in the magnetic field.

_____ **5.** Which of the following statements about ac generators is *not* true?
  a. The output current from the generator changes its direction at
  regular intervals.
  b. The emf alternates from positive to negative.
  c. A graph of Δemf versus Δt is similar to a sine curve.
  d. The frequency of alternating current is constant.

_____ **6.** Which of the following parts is used to convert alternating current
  into direct current?
  a. conducting or slip rings
  b. brushes
  c. split slip rings or commutators
  d. axis connectors

Electromagnetic Induction *continued*

_____ **7.** Motors are machines that convert
   a. mechanical energy into motion.
   b. electrical energy into mechanical energy.
   c. mechanical energy into quantum energy .
   d. none of the above

_____ **8.** Which of the following statements about back emf is not true?
   a. Since back emf is an induced current, the net current available to the motor increases.
   b. The faster the coil rotates, the greater the back emf becomes.
   c. The ac current applied to a motor causes the induced magnetic field to reverse regularly.
   d. The induced magnetic field is always repelled by the fixed magnetic field.

9. How do most commercial power plants obtain the mechanical energy that is needed to rotate the generator's turbine blades?

_____

_____

10. What is mutual inductance? What important electrical device is based on it?

_____

_____

_____

_____

Assessment

Electromagnetic Induction

Section Quiz: AC Circuits and Transformers

Write the letter of the correct answer in the space provided.

_____ **1.** The energy produced by an alternating current with a maximum value of I_{max} is not the same as that produced by a direct current of the same value because
   a. the alternating current reverses direction.
   b. the alternating current is less efficient.
   c. the alternating current is at its maximum value only for an instant.
   d. the alternating current moves electrons at a slower rate.

_____ **2.** The rms current is _____ the maximum current.
   a. less than
   b. equal to
   c. greater than
   d. the square root of

_____ **3.** When the dc current equals I_{max}, the power dissipated in an ac circuit equals _____ the power dissipated in a dc current.
   a. one-fourth
   b. one-half
   c. double
   d. quadruple

_____ **4.** Both rms current and rms emf values are _____ less than the maximum values.
   a. never
   b. seldom
   c. often
   d. always

_____ **5.** Which of the following equations shows the correct relationship between rms emf, rms current, and resistance?

   a. $\Delta V_{rms} = I_{rms} R$ **c.** $I_{rms} = \dfrac{\Delta V_{rms}}{R}$

   b. $R = \dfrac{\Delta V_{rms}}{I_{rms}}$ **d.** all of the above

_____ **6.** What is the name of the device that changes a small ac applied emf to a larger one or a large applied emf to a smaller one?
   a. converter
   b. transformer
   c. inverter
   d. exchanger

_____ **7.** If the secondary emf is less than the primary emf, which of the following is true about the number of turns in the coils?
   a. N_2 is equal to N_1.
   b. N_2 is greater than N_1.
   c. N_2 is less than N_1.
   d. N_1 is less than N_2.

_____ **8.** According to the transformer equation, $\Delta V_2 = \frac{N_2}{N_1} \Delta V_1$, the relationship between the potential difference and number of turns is
   a. direct.
   b. indirect.
   c. inverse.
   d. exponential.

9. How do power companies reduce the amount of power lost to resistive heating in long-distance transmission lines? Why? What device is used to achieve this goal?

_____

_____

_____

_____

10. An ac current with a maximum emf of 230 V is connected to a dryer whose resistance is 110 Ω. Calculate the rms potential difference (ΔV_{rms}). Find the rms current (I_{rms}) through the dryer. Find the maximum ac current (I_{max}) in the circuit.

Electromagnetic Induction

Section Quiz: Electromagnetic Waves

Write the letter of the correct answer in the space provided.

_____ 1. Electromagnetic waves consist of _____ electric and magnetic fields.
   a. oscillating
   b. constant
   c. parallel
   d. longitudinal

_____ 2. Which of the following statements correctly describes electromagnetic waves?
   a. The electric and magnetic fields are at right angles to each other.
   b. The electric and magnetic fields are at right angles to the direction that the wave is moving.
   c. Electromagnetic waves are transverse waves.
   d. all of the above

_____ 3. All electromagnetic waves are produced by _____ charges.
   a. stationary
   b. accelerating
   c. neutral
   d. constant

_____ 4. Which of the following reasons best supports the conclusion that electricity and magnetism are two aspects of a single electromagnetic force?
   a. The electromagnetic force obeys the inverse-square law.
   b. Both electric and magnetic fields produce forces on charged particles.
   c. The electromagnetic force is one of the four fundamental forces in the universe.
   d. none of the above

_____ 5. Where is energy stored in electromagnetic waves?
   a. in the oscillating electric and magnetic fields
   b. in the moving charged particles
   c. in the motion of the electromagnetic waves
   d. all of the above

Electromagnetic Induction *continued*

_____ **6.** Which of the following statements about the electromagnetic force is *not* true?
  a. The magnitude of the electromagnetic force is proportional to the charge.
  b. The electromagnetic force is proportional to the electric field strength.
  c. The electromagnetic force is proportional to the magnetic field strength.
  d. The electromagnetic force is inversely proportional to the charge.

_____ **7.** The energy transported by electromagnetic waves is called electromagnetic
  a. propagation.
  b. radiation.
  c. variation.
  d. deviation.

_____ **8.** What term refers to electromagnetic waves as particles?
  a. cathodes
  b. rays
  c. photons
  d. diodes

9. What is the electromagnetic spectrum?

_____

_____

_____

10. Choose one type of electromagnetic wave on the electromagnetic spectrum and describe its wavelength, frequency, and energy properties. Identify two applications of the electromagnetic wave.

_____

_____

_____

_____

Atomic Physics

Section Quiz: Quantization of Energy

Write the letter of the correct answer in the space provided.

_____ **1.** Which of the following phrases correctly describes a blackbody?
  a. object from which neither light nor matter escapes
  b. absorbs all radiation and emits no radiation
  c. emits all radiation and absorbs no radiation
  d. perfectly absorbs and emits all radiation

_____ **2.** Classical electromagnetic theory predicted that the energy radiated by
a blackbody would become infinite as the wavelength of the radiation
became shorter. What was the contradiction between observation and
this result called?
  a. the Compton shift
  b. the ultraviolet catastrophe
  c. the photoelectric effect
  d. the quantum theory

_____ **3.** Which of the following statements is true about the energy of a
quantum of radiation?
  a. Energy increases with wavelength.
  b. Energy increases with frequency.
  c. Energy increases with intensity.
  d. Energy increases with speed.

_____ **4.** What is the energy of a photon with a frequency of 5.45×10^{14} Hz?
($h = 6.63 \times 10^{-34}$ J•s)
  a. 3.61×10^{-19} J
  b. 3.61×10^{-34} J
  c. 3.65×10^{-40} J
  d. 1.22×10^{-48} J

_____ **5.** What is the frequency of a photon with an energy of 1.3×10^{-19} J?
($h = 6.63 \times 10^{-34}$ J•s)
  a. 8.6×10^{-20} Hz
  b. 1.5×10^{-6} Hz
  c. 2.0×10^{14} Hz
  d. 1.2×10^{52} Hz

_____ **6.** For a photoelectron to be emitted by a metal that is exposed to photons, the energy of the photons must be greater than what property of the metal?
  a. its threshold frequency
  b. its ionization energy
  c. its electronegativity
  d. its work function

_____ **7.** A metal with a work function of 3.5 eV is exposed to photons with an energy of 3.7 eV. What is the maximum kinetic energy of the emitted photoelectrons?
  a. 7.2 eV
  b. 3.7 eV
  c. 3.5 eV
  d. 0.2 eV

_____ **8.** Which of the following statements correctly describes the Compton shift that occurs when photons scatter from electrons?
  a. Electron momentum decreases as electrons scatter from photons.
  b. Photon wavelengths shorten as they gain energy from electrons.
  c. Photon wavelengths lengthen as they lose energy to electrons.
  d. Scattered photons interfere with each other at different angles.

9. Explain why the concept of quanta was required to make theoretical predictions for blackbody radiation match the experimental observations.

_____

_____

_____

_____

10. Photons with a frequency of 6.6×10^{14} Hz shine on the surface of a metal with a work function of 2.4 eV. What is the maximum kinetic energy of the emitted photoelectrons? ($h = 6.63 \times 10^{-34}$ J•s; 1 eV = 1.6×10^{-19} J)

Assessment

Atomic Physics

Section Quiz: Models of the Atom

Write the letter of the correct answer in the space provided.

_____ 1. What did Rutherford's experiment demonstrate?
   a. The atomic nucleus was a large spherical volume of positive charge.
   b. The atomic nucleus was a small, compact region of positive charge.
   c. The electrons were imbedded in the sphere of positive charge.
   d. The electrons moved around the nucleus like planets orbiting the sun.

_____ 2. What observation was *not* explained by the Rutherford model of the atom?
   a. the Compton shift
   b. emission spectra
   c. the photoelectric effect
   d. blackbody radiation

_____ 3. Which of the following statements is true about an emission spectrum?
   a. The wavelengths of the spectrum are the same for all elements.
   b. The lines of the spectrum are equally separated.
   c. It consists of dark lines in an otherwise continuous spectrum.
   d. It consists of narrow bright lines.

_____ 4. What type of spectrum is observed in the light from the sun and other stars?
   a. a continuous spectrum
   b. an emission spectrum
   c. an absorption spectrum
   d. an atomic spectrum

_____ 5. By what process does an electron in the Bohr model of the atom drop from a higher-energy level to a lower-energy level and emit a photon?
   a. line emission
   b. line absorption
   c. spontaneous emission
   d. spontaneous absorption

_____ 6. Which of the following is a feature of the Bohr model of the atom?
   a. Only specific electron orbits with given energies are stable.
   b. Electrons emit radiation continuously while orbiting the nucleus.
   c. Orbits of all possible energies are allowed for the atom's electrons.
   d. Electrons are located within a spherical region of positive charge.

_____ **7.** Which of the following statements correctly describes an atom's energy levels as they are depicted in an energy-level diagram?
   a. The energy levels are separated by equal amounts.
   b. The higher energy levels are separated by smaller amounts.
   c. The higher energy levels are separated by greater amounts.
   d. The energy levels are the same for all elements.

_____ **8.** A photon with an energy of 2.86 eV is absorbed by a hydrogen atom. Afterwards, three photons are spontaneously emitted. Which statement correctly describes the emitted photons?
   a. The emitted photons each have the same wavelengths.
   b. The photons are produced by a single electron energy-level transition.
   c. The sum of the emitted photon energies equals 2.86 eV.
   d. Three photons are always emitted when a 2.86 eV photon is absorbed.

9. Describe the Rutherford model of the atom, noting its strengths and weaknesses.

_____

_____

_____

_____

_____

_____

10. Describe the Bohr model of the atom, noting its strengths and weaknesses.

_____

_____

_____

_____

_____

_____

Assessment

Atomic Physics

Section Quiz: Quantum Mechanics

Write the letter of the correct answer in the space provided.

_____ 1. At what point do photons behave less like waves and more like particles?
   a. as the frequency of the photons increases
   b. as the wavelength of the photons increases
   c. as the intensity of the photons increases
   d. as the speed of the photons increases

_____ 2. Which of the following phenomena is the result of light's wave properties?
   a. the Compton shift
   b. two-slit interference
   c. the photoelectric effect
   d. momentum transfer

_____ 3. Which of the following experiments indicated that matter waves exist?
   a. the emission of electrons by a metal exposed to photons
   b. the diffraction of electrons by a single crystal
   c. the change in photon wavelength during scattering by electrons
   d. the spontaneous emission of photons by electron transitions in an atom

_____ 4. What is the momentum of a proton with a de Broglie wavelength of 6.63×10^{-9} m? ($h = 6.63 \times 10^{-34}$ J•s)
   a. 4.40×10^{-44} kg•m/s
   b. 3.33×10^{-34} kg•m/s
   c. 1.00×10^{-25} kg•m/s
   d. 3.00×10^{-17} kg•m/s

_____ 5. According to the matter-wave modification to the Bohr model of the atom, what do the orbits of electrons in an atom resemble?
   a. longitudinal waves
   b. probability waves
   c. traveling waves
   d. standing waves

| Atomic Physics *continued*

_____ 6. Which of the following statements correctly describes the results of simultaneous measurement of momentum and location for a particle?
   a. Accuracy of measurement decreases as the particle mass increases.
   b. Neither quantity can be measured with accuracy.
   c. The more accurately one quantity is measured, the less accurately the other quantity is known.
   d. Both quantities can be measured with infinite accuracy.

_____ 7. Which of the following indicates the greatest probability for an electron's position in the ground state of a hydrogen atom?
   a. the origin of the probability curve
   b. the middle of the rising part of the probability curve
   c. the middle of the peak of the probability curve
   d. the middle of the descending part of the probability curve

_____ 8. Which of the following does *not* describe the electron cloud for an electron in the ground state of a hydrogen atom?
   a. The electron cloud is spherically symmetrical.
   b. The electron cloud's density is close to zero near the nucleus.
   c. The electron cloud's density is greatest at the Bohr radius.
   d. The electron cloud's density is uniformly constant throughout.

9. Explain why the concept of photons provided a basis for the concept of matter waves.

_____

_____

_____

_____

_____

_____

_____

10. A muon has a mass of 1.88×10^{-28} kg. If it has a speed of 2.50×10^5 m/s, what is its de Broglie wavelength? ($h = 6.63 \times 10^{-34}$ J•s)

Assessment

Subatomic Physics

Section Quiz: The Nucleus

Write the letter of the correct answer in the space provided.

_____ 1. Which of the following statements correctly describes the relationship between the three nuclear quantities?
   a. Atomic number equals the mass number plus the neutron number.
   b. Mass number equals the atomic number plus the neutron number.
   c. Neutron number equals the mass number plus the atomic number.
   d. Mass number equals the neutron number minus the atomic number.

_____ 2. The mass number of a given isotope is equivalent to which of the following?
   a. the precise mass of the nucleus in atomic mass units
   b. the number of protons in the nucleus
   c. the number of neutrons in the nucleus
   d. the number of protons and neutrons in the nucleus

_____ 3. What is the energy associated with the mass of a subatomic particle called?
   a. internal energy
   b. rest energy
   c. nuclear energy
   d. binding energy

_____ 4. Which of the following statements correctly describes the necessary condition for a nucleus to be stable?
   a. The number of neutrons must increase more than the number of protons.
   b. The number of neutrons must equal the number of protons.
   c. The number of neutrons must increase less than the number of protons.
   d. Nuclear stability does not depend on the number of neutrons.

_____ 5. What force binds nucleons together in a nucleus?
   a. gravitational force
   b. Coulomb force
   c. strong force
   d. weak force

Subatomic Physics *continued*

_____ **6.** Why is the mass of a nucleus different than the sum of the masses of the nucleons that make up the nucleus?

    a. Some nucleons undergo beta decay, and so increase the nuclear mass.

    b. Some nucleons undergo beta decay, and so decrease the nuclear mass.

    c. Part of nucleon mass is converted to binding energy within the nucleus.

    d. The nucleus absorbs particles not bound to individual nucleons.

_____ **7.** The atomic mass of $^{52}_{24}$Cr is 51.940 511 u. What is the mass defect for this isotope? (atomic mass of $^1_1$H = 1.007 825 u; m_n = 1.008 665 u)

    a. 0.244 954 u

    b. 0.489 909 u

    c. 0.979 818 u

    d. 1.959 636 u

_____ **8.** The mass defect for a particular isotope is 0.466 769 u. What is the binding energy in MeV for the nucleus of this isotope? (c^2 = 931.49 MeV/u)

    a. 5.0109×10^{-4} MeV

    b. 5.0109 MeV

    c. 217.40 MeV

    d. 434.79 MeV

9. Explain why the number of neutrons required for stability by more-massive nuclei differs from the number of neutrons required for stability by less-massive nuclei.

_____

_____

_____

_____

_____

10. Calculate the binding energy of a $^{21}_{10}$Ne nucleus. (c^2 = 931.49 MeV/u; atomic mass of $^{21}_{10}$Ne = 20.993 841 u; atomic mass of $^1_1$H = 1.007 825 u; m_n = 1.008 665 u)

Name _____ Class _____ Date _____

Assessment

Subatomic Physics

Section Quiz: Nuclear Decay

Write the letter of the correct answer in the space provided.

_____ 1. When an alpha particle is emitted by a radioactive nucleus, which of the following changes occurs to the nucleus?
  a. The number of protons decreases by four.
  b. The number of protons decreases by two.
  c. The number of neutrons decreases by four.
  d. The number of nucleons decreases by two.

_____ 2. When a negatively-charged beta particle is emitted by a radioactive nucleus, which of the following changes occurs to the nucleus?
  a. The number of protons increases by one.
  b. The number of neutrons increases by one.
  c. The number of nucleons increases by one.
  d. The number of nucleons decreases by one.

_____ 3. Which radiation is the least penetrating?
  a. alpha
  b. beta
  c. gamma
  d. neutron

_____ 4. During beta decay, what particles are needed for momentum and energy to be conserved?
  a. electrons
  b. positrons
  c. neutrinos or antineutrinos
  d. neutrons

_____ 5. What does X represent in the nuclear decay formula: $^{232}_{90}\text{Th} \rightarrow X + ^{4}_{2}\text{He}$?
  a. $^{232}_{90}\text{Ra}$
  b. $^{228}_{88}\text{Ra}$
  c. $^{228}_{88}\text{Th}$
  d. $^{228}_{88}\text{He}$

_____ 6. What does X represent in the nuclear decay formula: $^{14}_{6}\text{C} \rightarrow X + ^{0}_{-1}e$?
  a. $^{14}_{7}\text{N}$
  b. $^{14}_{7}\text{C}$
  c. $^{14}_{8}\text{O}$
  d. $^{4}_{6}\text{N}$

Subatomic Physics *continued*

_____ **7.** How many half-lives does it take for a radioactive substance to decay to 25 percent of its original amount?

     a. 1

     b. 2

     c. 3

     d. 4

_____ **8.** A radioactive sample containing 24 g of material undergoes a decay process. After three half-lives have passed, how much of the original substance remains?

     a. 24 g

     b. 12 g

     c. 6 g

     d. 3 g

9. Give the rules of nuclear decay, and explain what properties must be conserved in a nuclear decay process.

_____

_____

_____

_____

_____

10. The half-life of astatine-218 is 1.6 s. If a sample of astatine-218 contains 1.6×10^{17} nuclei, how many nuclei, in becquerels, will decay each second?

Subatomic Physics

Section Quiz: Nuclear Reactions

Write the letter of the correct answer in the space provided.

_____ **1.** Which of the following processes correctly describes a nuclear fission reaction?
- **a.** Two light nuclei bind together to produce a heavier nucleus and energy.
- **b.** Two heavy nuclei bind together to produce heavier nuclei and energy.
- **c.** A heavy nucleus splits to produce lighter nuclei and energy.
- **d.** A heavy nucleus absorbs energy to produce lighter nuclei.

_____ **2.** Which of the following processes correctly describes a nuclear fusion reaction?
- **a.** Two light nuclei bind together to produce a heavier nucleus and energy.
- **b.** Two heavy nuclei bind together to produce heavier nuclei and energy.
- **c.** A heavy nucleus splits to produce lighter nuclei and energy.
- **d.** A heavy nucleus absorbs energy to produce lighter nuclei.

_____ **3.** What value must the mass number of an isotope be greater than for fission to occur?
- **a.** 26
- **b.** 58
- **c.** 83
- **d.** 112

_____ **4.** What property must a substance have if it is to control a fission chain reaction?
- **a.** It must absorb the daughter isotopes produced during the reaction.
- **b.** It must convert neutrons to protons during the reaction.
- **c.** It must emit additional neutrons to be used by the reaction.
- **d.** It must absorb a portion of the neutrons produced by the reaction.

_____ **5.** What type of particle is captured by other nuclei in a nuclear chain reaction, triggering additional fission events?
- **a.** electrons
- **b.** protons
- **c.** neutrons
- **d.** neutrinos

| Subatomic Physics *continued*

_____ **6.** What is a nuclear reactor?
  a. a system designed to increase the energy released from a chain reaction
  b. a system designed to start a fission chain reaction within heavy elements
  c. a system designed to start an uncontrolled, self-sustained chain reaction
  d. a system designed to maintain a controlled, self-sustained chain reaction

_____ **7.** What is the initial series of nuclear-fusion reactions that produces energy in the sun called?
  a. the hydrogen cycle
  b. the proton-proton cycle
  c. the hydrogen-helium cycle
  d. the deuterium-hydrogen cycle

_____ **8.** Which of the following is *not* an advantage of nuclear fusion reactors?
  a. Deuterium, the fuel used for fusion, is easily obtained from water.
  b. Fusion reactions produce few radioactive byproducts.
  c. The repulsive forces between deuterium nuclei are easily overcome.
  d. The fuel costs of a fusion reactor would be relatively low.

9. Explain why uranium-235 is used in fission reactions, but uranium-238 is not.

_____

_____

_____

_____

_____

10. Explain why fission reactions are much easier to harness than fusion reactions for the commercial production of energy.

_____

_____

_____

_____

_____

_____

Subatomic Physics

Section Quiz: Particle Physics

Write the letter of the correct answer in the space provided.

_____ **1.** What is the order of the fundamental interactions from strongest to weakest?
  a. strong, electromagnetic, weak, gravitational
  b. strong, electromagnetic, gravitational, weak
  c. strong, gravitational, electromagnetic, weak
  d. gravitational, strong, electromagnetic, weak

_____ **2.** Which of the following fundamental interactions are effective over a long range?
  a. the gravitational and strong interactions
  b. the electromagnetic and weak interactions
  c. the gravitational and electromagnetic interactions
  d. the strong and weak interactions

_____ **3.** Which of the following statements about leptons is correct?
  a. Leptons are made up of quarks.
  b. Leptons are fundamental particles.
  c. Leptons participate in all four fundamental interactions.
  d. Leptons participate only in the weak interaction.

_____ **4.** Which of the following statements about hadrons is correct?
  a. Hadrons are always made up of three quarks.
  b. Hadrons are fundamental particles.
  c. Hadrons participate in all four fundamental interactions.
  d. There are only six types of hadrons.

_____ **5.** What quarks are combined to form a neutron? ($u = +\frac{2}{3}e$ and $d = -\frac{1}{3}e$)
  a. two u and one d
  b. two d and one u
  c. two $\bar{u}$ and one $\bar{d}$
  d. two $\bar{d}$ and one $\bar{u}$

_____ **6.** Besides the field particles that mediate the various interactions, what basic types of particles, according to the standard model, make up all matter?
  a. three kinds of quarks and three kinds of leptons
  b. six kinds of quarks and three kinds of leptons
  c. six kinds of quarks and six kinds of leptons
  d. three kinds of quarks and six kinds of leptons

_____ **7.** According to the standard model, which field particles mediate the electromagnetic interaction?

  a. gluons

  b. bosons

  c. gravitons

  d. photons

_____ **8.** During the expansion of the universe after the big bang, which interaction separated from the other three fundamental interactions first?

  a. the gravitational interaction

  b. the strong interaction

  c. the electromagnetic interaction

  d. the weak interaction

9. Describe how leptons and hadrons are related, and how they differ.

_____

_____

_____

_____

10. One u quark and one $\bar{d}$ quark combine to form a pion (π^+). Show how the combination of these two quarks produces a charge of $+e$.

Answer Key

1 The Science of Physics

WHAT IS PHYSICS?

1. d   5. c
2. b   6. b
3. d   7. d
4. a   8. a
9. Answers will vary. Possible answers may describe the processes used in the investigation of a crime or car accident scene.
10. The first step is to define the system and three ways to summarize a complicated situation include the use of a diagram, computer simulation, or small-scale replica.

1 The Science of Physics

MEASUREMENTS IN EXPERIMENTS

1. a
2. d
3. c

Solution

$674.3 \text{ mm} \times 10^2 / 100$
$= \boxed{6.743 \times 10^2 \text{ mm}}$

4. d
    (a) $0.000\ 005\ 823 \text{ μg} \times 10^{-6} / 0.000\ 001 = 5.823 \times 10^{-6} \text{ μg}$
    (b) $5.823 \times 10^{-6} \text{ μg} \times 10^{-6} \text{ g} / 1 \text{ μg} = 5.823 \times 10^{-12} \text{ g}$
    (c) $5.823 \times 10^{-6} \text{ μg} \times 10^{-3} \text{ mg} / 1 \text{ μg} = 5.823 \times 10^{-9} \text{ mg}$
    (d) Correct, since a–c are all correct responses.
5. b

Solution

$1.673 \times 10^{-27} \text{ kg} \times 10^3 \text{ g} / 1 \text{ kg} =$
$\boxed{1.673 \times 10^{-24} \text{ g}}$

6. a
7. c
8. b
9. A measurement indicates precision, since its significant figures include all digits that are actually measured plus one estimated digit.

10. *Given*
    Length = 15.23 m
    Width = 8.7 m
    Solution
    Area = 15.23 m × 8.7 m = 132.501 m, which rounds to 130 m (only two significant digits are permitted) and is $\boxed{1.3 \times 10^2 \text{ m}}$ in scientific notation.

1 The Science of Physics

THE LANGUAGE OF PHYSICS

1. c
2. d
3. a

Solution
Based on Table 1 and Graph 1, substance B shows a direct relationship between temperature and time with an average ΔT of 0.7°C for each 10 s time interval. Therefore, "a" is the correct answer.

4. a
5. b
6. c
7. b
8. d
9. Equations describe the relationship between variables, and enable one to produce graphs and to make predictions.
10. Order-of-magnitude calculations provide estimates: (1) that permit a scientist to judge whether an answer from a more exacting procedure is correct; and (2) in situations in which little information is given.

2 Motion in One Dimension

DISPLACEMENT AND VELOCITY

1. b   5. c
2. a   6. d
3. b   7. c
4. c   8. d
9. The first time the dog is at 1.0 m it rests for 2.0 seconds. The second time it is at 1.0 m it quickly changes direction from south to north.

10. 0.10 m/s south

Given

from the graph, $d_i = 3.0$ m and $d_f = 4.0$ m, $t_f = 10.0$ s, and $t_i = 0.0$ s

Solution

$$v_{avg} = \frac{d_f - d_i}{t_f - t_i} =$$

$$\frac{[(+4.0 \text{ m}) - (+3.0 \text{ m})]}{(10.0 \text{ s} - 0.0 \text{ s})} =$$

$\boxed{+0.10 \text{ m/s to the south}}$

2 Motion in One Dimension

ACCELERATION

1. d	**5.** a
2. b	**6.** d
3. c	**7.** b
4. c	**8.** d

9. CD, DE, EF, and FG. The displacement of the jogger with a constant acceleration is equal to the average velocity during a time interval multiplied by the time interval. Because the time intervals are equal, the displacements are in the same order as decreasing average velocities. By inspection the average velocity decreases in the order CD, DE, EF, and FG.

10. 2.6 m

Given

$v_i = 0.25$ m/s
$a = 0.40$ m/s^2
$\Delta t = 3.0$ s

Solution

$\Delta x = v_i \Delta t + \frac{1}{2}a(\Delta t)^2 =$

$(0.25 \text{ m/s})(3.0 \text{ s}) + \frac{(0.40 \text{ m/s}^2)(3.0 \text{ s})^2}{2}$

$= \boxed{2.6 \text{ m}}$

2 Motion in One Dimension

FALLING OBJECTS

1. d	**5.** c
2. c	**6.** b
3. c	**7.** d
4. c	**8.** c

9.

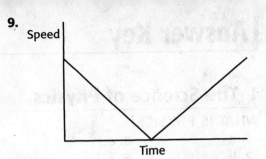

Speed / Time

10. -9.3 m/s

Given

$v_i = -3.0$ m/s^2
$a = -g = -9.81$ m/s^2
$\Delta x = -4.0$ m

Unknown

v_f

Solution

$v_f^2 = v_i^2 + 2a\Delta x$

$v_f = \pm\sqrt{v_i^2 + 2a\Delta x} =$

$\pm\sqrt{(-3.0 \text{ m/s})^2 + (2)(-9.81 \text{ m/s}^2)(-4.0 \text{ m})}$

$= \boxed{-9.3 \text{ m/s}}$

3 Two-Dimensional Motion and Vectors

INTRODUCTION TO VECTORS

1. c	**5.** d
2. d	**6.** c
3. b	**7.** b
4. b	**8.** b

9. A vector quantity is described by magnitude and direction; a scalar quantity is described only by magnitude.

10.

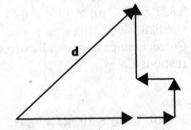

The resultant displacement, **d**, is about 4.2 blocks 45° north of east.

3 Two-Dimensional Motion and Vectors

VECTOR OPERATIONS

1. c	5. a
2. d	6. c
3. d	7. b
4. a	8. c

9. The vectors must be resolved into components. Then the horizontal component must be added and the vertical components must be added. The two perpendicular resultants can now be added using the Pythagorean theorem.

10. 220 m

Given

$d = 310$ m 45° east of south

Solution

$d_y = d\cos\theta = (310 \text{ m})(\cos 45°) =$

$\boxed{220 \text{ m}}$

3 Two-Dimensional Motion and Vectors

PROJECTILE MOTION

1. d	5. a
2. a	6. a
3. c	7. d
4. a	8. b

9. If the object is launched and caught from the same initial vertical position, its vertical displacement is zero because it was moving first up and then down the same distance while moving horizontally during its flight.

10. 2.6 m/s

Given

$\Delta y = -3.0$ m

$\Delta x = 2.0$ m

Solution

$\Delta y = \frac{1}{2}a_y(\Delta t)^2$

$\Delta t = \sqrt{\dfrac{2\Delta y}{a_y}} = \sqrt{\dfrac{2\Delta y}{-g}}$

$v_x = v_{avg,\,x} = \dfrac{\Delta x}{\Delta t} = \Delta x\sqrt{\dfrac{-g}{2\Delta y}} =$

$(2.0 \text{ m})\sqrt{\dfrac{(-9.81 \text{ m/s}^2)}{2(-3.0 \text{ m})}} = \boxed{2.6 \text{ m/s}}$

3 Two-Dimensional Motion and Vectors

RELATIVE MOTION

1. a	5. c
2. a	6. b
3. d	7. a
4. d	8. c

9. The dummy's motion relative to the camera would be zero. So, the dummy would appear to be at rest with respect to the camera.

10. 0.58 m/s

Given

$v_{dc} = +0.50$ m/s due north

$v_{cb} = +0.30$ m/s due east

Solution

$v_{db} = v_{dc} + v_{cb}$

$(v_{db})^2 = (v_{dc})^2 + (v_{cb})^2$

$v_{db} = \sqrt{(v_{dc})^2 + (v_{cb})^2} =$

$\sqrt{(0.50 \text{ m/s})^2 + (0.30 \text{ m/s})^2}$

$v_{db} = \boxed{0.58 \text{ m/s}}$

4 Forces and the Laws of Motion

CHANGES IN MOTION

1. b	5. d
2. b	6. a
3. d	7. c
4. d	8. d

9. 1. Identify the forces acting on the body and the direction of the forces.
   2. Draw a diagram to represent the isolated object.
   3. Draw and label force vectors for all the external forces acting on the object.

10.

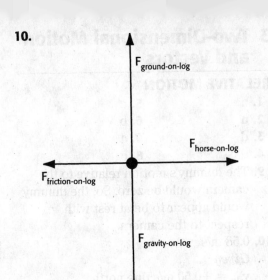

$F_{\text{ground-on-log}}$

$F_{\text{horse-on-log}}$

$F_{\text{friction-on-log}}$

$F_{\text{gravity-on-log}}$

4 Forces and the Laws of Motion

NEWTON'S FIRST LAW

1. d   **5.** d
2. a   **6.** b
3. d   **7.** d
4. c   **8.** c
9. The net force has not changed because the pillow is still at rest.
10. 215 N

Given

$F_g = 165$ N
$\theta = 50.0°$

Solution

$F_{net,y} = \Sigma F_y = F_g - F_{applied,\, y} = 0$
$F_{applied,\, y} = F_g$

$F_{applied} = \dfrac{F_{applied,\, y}}{\sin \theta} = \dfrac{F_g}{\sin \theta}$

$= \dfrac{(165 \text{ N})}{(\sin 50.0°)} = \boxed{215 \text{ N}}$

4 Forces and the Laws of Motion

NEWTON'S SECOND AND THIRD LAWS

1. c   **5.** c
2. b   **6.** d
3. d   **7.** b
4. d   **8.** b

9. In the action-reaction pair involving the book and Earth, Earth exerts a downward force on the book equal to the force of gravity on the book. In the action-reaction pair involving the book and table, the book exerts a downward force on the table equal to the force of gravity acting on the book. The table exerts an equal magnitude force upward the book. Because the book has a downward force and upward force acting on it which are equal in magnitude, the net force on the book is zero. Therefore, the book is in equilibrium.

10. 0.8 m/s^2 in the direction of the child

Solution
$F_{net} = \Sigma F_x = F_1 - F_2 =$
$16.3 \text{ N} - 15.8 \text{ N} = 0.5 \text{ N}$
$F_{net} = ma$

$a = \dfrac{F_{net}}{m} = \dfrac{0.5 \text{ N}}{0.62 \text{ kg}} =$

$\boxed{0.8 \text{ m/s}^2 \text{ in the direction of the child}}$

4 Forces and the Laws of Motion

EVERYDAY FORCES

1. d   **5.** b
2. d   **6.** b
3. c   **7.** d
4. d   **8.** d

9. Static friction and kinetic friction are measured using the same surfaces in contact and the normal force remains the same.

10. 43 N

Given

$F_g = 18$ N
$\mu_k = 0.42$

Solution

$F_{net,\, y} = \Sigma F_y = F_f - F_g = 0$
$F_f = F_g$
$F_{net,\, x} = \Sigma F_x = F_n - F_{applied} = 0$
$F_{applied} = F_n$
$F_f = \mu_k F_n$

$F_n = \dfrac{F_f}{\mu_k}$

$F_{applied} = \dfrac{F_g}{\mu_k} = \dfrac{18 \text{ N}}{0.42} = \boxed{43 \text{ N}}$

5 Work and Energy

WORK

1. d
2. c
3. b
4. c
5. a
6. b
7. b
8. d

9. While lifting the block, the worker does positive work on the block while gravity does negative work on the block. The net work while lifting the block is positive. When the worker is holding the block, no forces do work on the block and no net work is done on the block. While lowering the block, the worker does positive work while gravity does negative work on the block. The net work on the block while it is lowered is negative. The total net work on the block is zero because the net displacement is zero.

10. 99 J

Given

$d = 3.0$ m
$F_{child} = 55$ N
$\theta = 35°$
$F_k = -12$ N

Solution

$W_{net} = F_{net}d = (F_{child}\cos\theta + F_k)d =$
  $[(55 \text{ N})(\cos 35°) + (-12 \text{ N})](3.0 \text{ m})$
  $= \boxed{99 \text{ J}}$

5 Work and Energy

ENERGY

1. a
2. c
3. d
4. c
5. b
6. b
7. d
8. c

9. The bocce ball has more kinetic energy. Kinetic energy depends on both mass and velocity. However, kinetic energy is more strongly dependent on velocity because the velocity term is squared in the equation for kinetic energy: $KE = (1/2)mv^2$.

10. $KE_i = 1.1 \times 10^5$ J; $KE_f = 8.5 \times 10^4$ J

Given

$m = 1.0 \times 10^3$ kg
$v_i = 15$ m/s^2
$W_{net} = -25$ kJ $= -2.5 \times 10^4$ J

Solution

$KE_i = \left(\frac{1}{2}\right)mv_i^2 = \left(\frac{1}{2}\right)(1.0 \times 10^3 \text{ kg})$
  $(15 \text{ m/s})^2 = 1.1 \times 10^5$ J
$W_{net} = \Delta KE = KE_f - KE_i$
$KE_f = KE_i + W_{net} = (1.1 \times 10^5 \text{ J}) +$
  $(-2.5 \times 10^4 \text{ J}) = \boxed{8.5 \times 10^4 \text{ J}}$

5 Work and Energy

CONSERVATION OF ENERGY

1. d
2. c
3. d
4. a
5. d
6. b
7. d
8. c

9. When the ball is first thrown, the ball has some kinetic energy and some gravitational potential energy. As the ball rises, the kinetic energy is transferred to gravitational potential energy. At the peak, all the energy is potential energy. As the ball falls, the potential energy is transferred to kinetic energy. When the ball hits the ground, all the energy is kinetic energy. Mechanical energy is conserved throughout the flight of the ball.

10. 5.8 m/s

Given

$m = 5.7 \times 10^{-2}$ kg
$v_i = 2.0$ m/s
$h_i = 1.5$ m
$h_f = 0$ m
$g = 9.81$ m/s^2

Solution

$ME_i = ME_f$
$\frac{1}{2}mv_i^2 + mgh_i = \frac{1}{2}mv_f^2 + mgh_f$
$v_f^2 = \dfrac{2\left(\frac{1}{2}mv_i^2 + mgh_i - mgh_f\right)}{m}$
$v_f = \sqrt{2\left(\frac{1}{2}v_i^2 + gh_i - gh_f\right)}$

$v_f =$
$\sqrt{2\left[\left(\frac{1}{2}\right)(2.0 \text{ m/s})^2 + (9.81 \text{ m/s}^2)(1.5 \text{ m}) - (9.81 \text{ m/s}^2)(0 \text{ m})\right]}$

$v_f = \boxed{5.8 \text{ m/s}}$

5 Work and Energy

POWER

1. d
2. d
3. c
4. c
5. c
6. a
7. b
8. d

9. Power measures the amount of energy that is transferred from one object to another or transformed to other forms of energy in a given time interval.

10. 220 kW

   Given

   $F = 29 \text{ kN} = 2.9 \times 10^4 \text{ N}$

   $v = 7.5 \text{ m/s}$

   Solution

   $P = Fv = (2.9 \times 10^4 \text{ N})(7.5 \text{ m/s}) =$
   $2.2 \times 10^5 \text{ W} = \boxed{220 \text{ kW}}$

6 Momentum and Collisions

MOMENTUM AND IMPULSE

1. b
2. c
3. a
4. c
5. b
6. c
7. c
8. d

9. Impulse is the product of the force acting on an object and the time interval in which the force acts on an object. The impulse-momentum theorem states that the impulse on an object is equal to the change in the object's momentum.

10. $9.0 \times 10^6 \text{ kg} \bullet \text{m/s}$

   Given

   $m = 1.0 \times 10^4 \text{ kg}$

   $v_i = 1.2 \times 10^3 \text{ m/s}$

   $F = -25 \text{ kN} = 2.5 \times 10^4 \text{ N}$

   $\Delta t = 2.0 \text{ min} = 120 \text{ s}$

   Solution

   $p_i = mv_i =$
   $(1.0 \times 10^4 \text{ kg})(1.2 \times 10^3 \text{ m/s}) =$
   $1.2 \times 10^7 \text{ kg} \bullet \text{m/s}$

   $F\Delta t = \Delta p = p_f - p_i$

   $p_f = F\Delta t + p_i$

   $p_f = (-2.5 \times 10^4 \text{ N})(120 \text{ s}) +$
   $(1.2 \times 10^7 \text{ kg} \bullet \text{m/s}) =$
   $\boxed{9.0 \times 10^6 \text{ kg} \bullet \text{m/s}}$

6 Momentum and Collisions

CONSERVATION OF MOMENTUM

1. d
2. c
3. b
4. d
5. d
6. c
7. d
8. c

9. The magnitude of the momentum of the slower ball increases, while the magnitude of the momentum of the faster ball decreases by the same amount. Both balls reverse direction in the collision. The total momentum of the system does not change.

10. 1.9 m/s

   Given

   $m_1 = 55 \text{ kg}$

   $v_{1,i} = 2.0 \text{ m/s}$

   $m_2 = 2.0 \text{ kg}$

   $v_{2,i} = 0 \text{ m/s}$

   $v_{1,f} = v_{2,f}$

   Solution

   $m_1 v_{1,i} + m_2 v_{2,i} = m_1 v_{1,f} + m_2 v_{2,f}$

   $m_1 v_{1,i} = m_1 v_{1,f} + m_2 v_{1,f}$

   $m_1 v_{1,i} = (m_1 + m_2) v_{1,f}$

   $v_{1,f} = \dfrac{m_1 v_{1,i}}{(m_1 + m_2)}$

   $v_{1,f} = \dfrac{(55 \text{ kg})(2.0 \text{ m/s})}{(55 \text{ kg} + 2.0 \text{ kg})} = \boxed{1.9 \text{ m/s}}$

6 Momentum and Collisions

ELASTIC AND INELASTIC COLLISIONS

1. d
2. b
3. a
4. b
5. d
6. a
7. b
8. d

9. An inelastic collision is any collision in which some kinetic energy is converted to other forms of energy so that the total kinetic energy is not conserved. A perfectly inelastic collision is a special case in which the objects in the collision stick together and move as a single object after the collision.

10. 1.2 m/s to the right

   Given

   $m_1 = 0.16 \text{ kg}$

   $v_{1,i} = 1.2 \text{ m/s}$

   $m_2 = 0.16 \text{ kg}$

   $v_{2,i} = -0.85 \text{ m/s}$

   $v_{1f} = -0.85 \text{ m/s}$

Solution

$$m_1v_{1,i} + m_2v_{2,i} = m_1v_{1f} + m_2v_{2f}$$

$$v_{2,f} = \frac{(m_1v_{1,i} + m_2v_{2,i} - m_1v_{1f})}{m_2}$$

$$v_{2,f} =$$

$$\frac{(0.16 \text{ kg})(1.2 \text{ m/s}) + (0.16 \text{ kg})(-0.85 \text{ m/s}) - (0.16 \text{ kg})(-0.85 \text{ m/s})}{0.16 \text{ kg}}$$

$$v_{2,f} = \boxed{1.2 \text{ m/s to the right}}$$

$$KE_i = \frac{1}{2}mv_{1,i}^2 + \frac{1}{2}mv_{2,i}^2 =$$

$$\frac{1}{2}(0.16 \text{ kg})(1.2 \text{ m/s})^2 +$$

$$\frac{1}{2}(0.16 \text{ kg})(-0.85 \text{ m/s})^2$$

$$KE_i = 0.12 \text{ J} + 0.058 \text{ J} = 0.18 \text{ J}$$

$$KE_f = \frac{1}{2}mv^2_{1f} + \frac{1}{2}mv^2_{2f} =$$

$$\frac{1}{2}(0.16 \text{ kg})(-0.85 \text{ m/s})^2 +$$

$$\left(\frac{1}{2}\right)(0.16 \text{ kg})(1.2 \text{ m/s})^2$$

$$KE_f = 0.058 \text{ J} + 0.12 \text{ J} = 0.18 \text{ J}$$

$$KE_i = KE_f$$

7 Circular Motion and Gravitation

CIRCULAR MOTION

1. b
2. c
3. a
4. b
5. c
6. d
7. b
8. d

9. Friction between the car's tires and the road is the centripetal force that causes the car to move along a curved or circular path. Passengers in the car tend to lean or slide toward the outside of the turn because their inertia causes them to tend toward moving in a straight-line path.

10. $a_c = 0.83 \text{ m/s}^2$; $F_c = 1.1 \times 10^3 \text{ N}$

Given

$v_t = 2.5 \text{ m/s}$

$r = 7.5 \text{ m}$

$m = 1.3 \times 10^3 \text{ kg}$

Solution

$$a_c = \frac{v_t^2}{r} = \frac{(2.5 \text{ m/s})^2}{7.5 \text{ m}} = \boxed{0.83 \text{ m/s}^2}$$

$$F_c = \frac{mv_t^2}{r} = \frac{(1.3 \times 10^3 \text{ kg})(2.5 \text{ m/s})^2}{7.5 \text{ m}}$$

$$= \boxed{1.1 \times 10^3 \text{ N}}$$

7 Circular Motion and Gravitation

NEWTON'S LAW OF UNIVERSAL GRAVITATION

1. b
2. c
3. a
4. c
5. d
6. a
7. d
8. d

9. Weight is the product of mass and gravitational field strength. An astronaut weighs less on the moon than on Earth because the gravitational field strength at the moon's surface is less than the gravitational field strength on Earth's surface.

10. $1.4 \times 10^2 \text{ N}$

Given

$m_1 = 7.35 \times 10^{22} \text{ kg}$

$m_2 = 85 \text{ kg}$

$r = 1.74 \times 10^6 \text{ m}$

$G = 6.673 \times 10^{-11} \text{ N} \bullet \text{m}^2/\text{kg}^2$

Solution

$$F_g = G\frac{m_1m_2}{r^2} =$$

$$(6.673 \times 10^{-11} \text{ N} \bullet \text{m}^2/\text{kg}^2) \times$$

$$\frac{(7.35 \times 10^{22} \text{ kg})(85 \text{ kg})}{(1.74 \times 10^6 \text{ m})^2} =$$

$$\boxed{1.4 \times 10^2 \text{ N}}$$

7 Circular Motion and Gravitation

MOTION IN SPACE

1. c
2. d
3. b
4. c
5. a
6. d
7. c
8. d

9. The astronaut is in free fall at the same rate of acceleration as his or her surroundings.

10. $r = 5240 \text{ s}$; $V_t = 7820 \text{ m/s}$

Given

$altitude = 139 \text{ km} = 1.39 \times 10^5 \text{ m}$

$r_E = 1.74 \times 10^6 \text{ m}$

$m_E = 5.97 \times 10^{24} \text{ kg}$

$G = 6.673 \times 10^{-11} \text{ N} \bullet \text{m}^2/\text{kg}^2$

Solution

$r = altitude + r_E = (1.39 \times 10^5 \text{ m}) +$

$(6.38 \times 10^6 \text{ m}) = 6.52 \times 10^6 \text{ m}$

$$T = 2\pi\sqrt{\frac{r^3}{Gm}} =$$

$$2\pi\sqrt{\frac{(6.52 \times 10^6 \text{ m})^3}{(6.673 \times 10^{-11} \text{ N}\bullet\text{m}^2/\text{kg}^2)(5.97 \times 10^{24} \text{ kg})}}$$

$$= 5240 \text{ s}$$

$$v_t = \sqrt{G\frac{m}{r}} =$$

$$(6.673 \times 10^{-11} \text{ N}\bullet\text{m}^2/\text{kg}^2) \times$$

$$\frac{(5.97 \times 10^{24} \text{ kg})}{(6.52 \times 10^6 \text{ m})} = \boxed{7820 \text{ m/s}}$$

7 Circular Motion and Gravitation

TORQUE AND SIMPLE MACHINES

1. d
2. a
3. d
4. b
5. b
6. c
7. b
8. d

9. In order for a machine to have 100% efficiency, the machine would have to be totally frictionless. Because any real machine has some friction, some of the energy input into a real machine is converted to nonmechanical forms of energy. As a result, the work output is always less than the work input.

10. 97 cm from the right end

Given

$m_1 = 35 \text{ kg}$

$m_2 = 85 \text{ kg}$

$d_1 = -1.50 \text{ m} + 0.20 \text{ m} = -1.30 \text{ m}$

$g = -9.81 \text{ m/s}^2$

Solution

$F_1 = m_1 g = (35 \text{ kg})(-9.81 \text{ m/s}^2) = -340 \text{ N}$

$F_2 = m_2 g = (85 \text{ kg})(-9.81 \text{ m/s2}) = -830 \text{ N}$

$\tau_{net} = F_1 d_1 + F_2 d_2 = 0$

$F_2 d_2 = -F_1 d_1$

$d_2 = \frac{-F_1 d_1}{F_2} = \frac{-(-340 \text{ N})(-1.30 \text{ m})}{(-830 \text{ N})}$

$= 0.53 \text{ m}$

$d = 1.50 \text{ m} - 0.53 \text{ m} = 0.97 \text{ m} =$

$\boxed{97 \text{ cm from the right end}}$

8 Fluid Mechanics

FLUIDS AND BUOYANT FORCE

1. a
2. d
3. a

4. d

Given

weight of displaced water = F_g = 10.0 N

5. b
6. c
7. c
8. a

9. Fluids do not possess definite shape, because the atoms or molecules in the fluid are free to move past each other. Ice is a solid in which the water molecules are bound together in a crystalline arrangement that prevents their moving past each other. Ice therefore has definite a shape, and does not flow. The molecules in liquid water or steam are able to move past each other, so that liquid water or steam flows and has no definite shape, and therefore is a fluid.

10. $2.2 \times 10^{-2} \text{ N}$

The metal is more dense than the salt water, so it is completely submerged. The volume of the displaced salt water (V_{sw}) equals the volume of the metal (V_m).

Given

$\rho_{sw} = 1.025 \times 10^3 \text{ kg/m}^3$

$\ell = 1.3 \text{ cm}$

$g = 9.81 \text{ m/s}^2$

Solution

$F_b = \rho_{sw} V_{sw} g = \rho_{sw} V_m g = \rho_{sw} \ell^3 g$

$F_b = (1.025 \times 10^3 \text{ kg/m}^3)(1.3 \text{ cm})^3$

$(9.81 \text{ m/s}^2) \times \left(\frac{1 \text{m}}{100 \text{ cm}}\right)^3$

$F_b = \boxed{2.2 \times 10^{-2} \text{ N}}$

8 Fluid Mechanics

FLUID PRESSURE

1. c
2. a
3. c

Given

$P = 8.0 \times 10^4 \text{ Pa}$

$A = 1.0 \times 10^{-2} \text{ m}^2$

Solution

$F = PA = (8.0 \times 10^4 \text{ Pa})(1.0 \times 10^{-2} \text{ m}^2)$

$= 8.0 \times 10^2 \text{ N}$

4. b

5. d

Given

$F_1 = 580$ N

$A_1 = 2.0$ m^2

$F_2 = 2900$ N

Solution

$A_2 = \left(\dfrac{A_1}{F_1}\right)F_2 = \left(\dfrac{2.0 \text{ m}^2}{580 \text{ N}}\right)(2900 \text{ N}) = 1.0 \times 10^1$ m^2

6. c

Given

$P = 3.03 \times 10^5$ Pa

$P_0 = 1.01 \times 10^5$ Pa

Solution

$\rho h g = P - P_0 = 3.03 \times 10^5$ Pa $- 1.01 \times 10^5$ Pa $= 2.02 \times 10^5$ Pa

7. a

8. c

9. Although a smaller force is needed to overcome a larger force, the concept of mechanical advantage (energy conservation) states that the amount of work done must be at least equal to that done in lifting a mass a certain distance. In a hydraulic lift, the pressure is uniform throughout the connecting fluid according to Pascal's principle. This means that the applied force acts on a smaller area than the weight being lifted, so the applied force must be smaller than the lifted weight. Because the applied force is smaller than the weight it is lifting, the distance through which the applied force acts must be greater than the distance the weight is raised.

10. 42 N

Given

$m_1 = 1.5 \times 10^3$ kg

$r_1 = 1.5$ m

$r_2 = 8.0$ cm

$g = 9.81$ m/s2

Solution

$P_1 = P_2$

$\dfrac{F_1}{A_1} = \dfrac{F_2}{A_2}$

$F_2 = \dfrac{F_1 A_2}{A_1} = \dfrac{(mg)(\pi(r_2)^2)}{(\pi(r_1)^2)} = (mg)\left(\dfrac{r_2}{r_1}\right)^2$

$F_2 = (1.5 \times 10^3 \text{ kg})(9.81 \text{ m/s}^2)$

$\left(\dfrac{8.0 \text{ cm}}{1.5 \text{ m}}\right)^2 \times \left(\dfrac{1 \text{ m}}{100 \text{ cm}}\right)^2 = \boxed{42 \text{ N}}$

8 Fluid Mechanics

FLUIDS IN MOTION

1. b

2. c

3. a

4. b

5. c

Given

flow rate $= 5.0$ m^3/s

$v = 7.5$ m/s

Solution

$A = \dfrac{\text{flow rate}}{v} = \dfrac{5.0 \text{ m}^3/\text{s}}{7.5 \text{ m/s}} = 0.67$ m^2

6. d

Given

$A_1 = 2.00$ m^2

$v_1 = 3.5$ m/s

$A_2 = 0.50$ m^2

Solution

$v_2 = \dfrac{A_1 v_1}{A_2} = \dfrac{(2.00 \text{ m}^2)(3.5 \text{ m/s})}{0.50 \text{ m}^2} = 14$ m/s

7. d

8. a

9. From the continuity equation, the speed of a fluid in a narrowing pipe increases. This increase in speed indicates that the fluid is being accelerated forward by a net force, and that this force arises from a difference in pressure within the fluid. Because the force is in the forward direction, the pressure at the front of the fluid must be lower than the pressure behind it. Thus pressure within the fluid decreases as the fluid's speed increases.

10. As the airplane moves forward with greater speed, the flow of air around the airplane wing reduces the pressure of the air. The wing is shaped so that air flows more rapidly over the wing than under it, and therefore causes the pressure above the wing to be lower than the pressure below it. The greater pressure under the wing raises the wing, causing the airplane to lift.

9 Heat

TEMPERATURE AND THERMAL EQUILIBRIUM

1. b
2. d
3. b
4. c
5. a
6. c

Given

$T = 235$ K

Solution

$T_C = T - 273.15 = (235 - 273.15)°C$
$\quad = -38°C$

7. d
8. a
9. Energy is added to a water molecule, causing its kinetic energy to increase. The kinetic energy can be distributed through the molecule's translational motion in all directions, through its rotational motion around its center of mass, and through vibrational motion along the bonds between the oxygen and hydrogen atoms. As the motion of the molecule and other molecules becomes greater, they convey more energy during collisions to other forms of matter, increasing their kinetic energies. These increases in energy account for and are measured as increases in temperature.
10. 97.9° F

Given

$T = 309.7$ K

Solution

$T = T_C + 273.15$

$T_C = T - 273.15$

$T_F = \dfrac{9}{5} T_C + 32.0 =$

$\dfrac{9}{5}(T - 273.15) + 32.0$

$T_F = \left(\dfrac{9}{5}(309.7 - 273.15) + 32.0\right)°F =$

$\left(\dfrac{9}{5}(36.6) + 32.0\right)°F =$

$(65.9 + 32.0)°F = \boxed{97.9°F}$

9 Heat

DEFINING HEAT

1. c
2. b
3. a
4. c
5. d

Given

$T_c = 37°C$

$T_w = 15°C$

Solution

$\Delta T = 37°C - 15°C = 22°C =$
$\quad 29°C - 7°C$

6. c
7. b
8. c

Given

$\Delta U = 18$ J

Solution

$\Delta PE = 0$ J; $KE_f = 0$ J;

$\Delta PE + \Delta KE + \Delta U = 0 = -KE_i + \Delta U$

$KE_i = \Delta U = 18$ J

9. Energy is transferred as heat from the particles with higher kinetic energies, and thus higher temperature, to particles with lower kinetic energies, and lower temperature, through collisions. In most cases, energy is given up from the particles with higher energies to those with lower energies. Overall, this causes the temperature of the object originally at higher temperature to decrease, and the temperature of the originally lower temperature object to increase. When the average kinetic energies of particles in both objects are the same, the amount of energy transferred from the first object to the second is the same as the energy transferred from the second object to the first. At this point, no net energy is exchanged, and the objects have the same temperatures. They are then in thermal equilibrium.
10. 401 m

Given

$\Delta U = 1770$ J

$m = 0.450$ kg

$g = 9.81$ m/s^2

Solution

$\Delta PE + \Delta KE + \Delta U = 0$

The kinetic energy increases with the decrease in potential energy, and then decreases with the increase in the internal energy of the water. Thus, the net change in kinetic energy is zero. Assuming final potential energy has a value of zero, the change in the internal energy equals:

$0 - PE_i + \Delta U = 0$

$\Delta U = PE_i = mgh$

$h = \dfrac{\Delta U}{mg} = \dfrac{1770 \text{ J}}{(0.450 \text{ kg})(9.81 \text{ m/s}^2)} =$

$\boxed{401 \text{ m}}$

9 Heat

CHANGES IN TEMPERATURE AND PHASE

1. b
2. c
3. a
4. d

Given

$Q = 3.6 \times 10^3 \text{ J}$

$m = 0.25 \text{ kg}$

$c_p = 360 \text{ J/kg} \cdot {}^\circ\text{C}$

Solution

$\Delta T = \dfrac{Q}{C_p m} = \dfrac{3.6 \times 10^3 \text{ J}}{(360 \text{ J/kg} \cdot {}^\circ\text{C})(0.25 \text{ kg})}$

$= 4.0 \times 10^1 \, {}^\circ\text{C}$

5. a
6. c
7. b
8. a
9. Water has a much larger specific heat capacity than does air. Therefore, for a given amount of energy provided to both by sunlight, the temperature of each kilogram of air rises more than does the temperature of a kilogram of water. Over time, the air reaches a high temperature and becomes hot while the water's temperature has not increased that much, and therefore still remains cool.

10. $5.2 \times 10^2 \text{ J/kg} \cdot {}^\circ\text{C}$

Given

$T_m = 93.0 {}^\circ\text{C}$

$T_w = 25.0 {}^\circ\text{C}$

$T_f = 29.0 {}^\circ\text{C}$

$m_m = 7.50 \times 10-2 \text{ kg}$

$m_w = 0.150 \text{ kg}$

$c_{p,w} = 4186 \text{ J/kg} \cdot {}^\circ\text{C}$

Solution

$Q_w = -Q_m$

$C_{p,w} m_w \Delta T_w = -C_{pm} m_m \Delta T_m$

$C_{p,w} m_w (T_f - T_m) =$
$\quad -C_{p,m} m_m (T_f - T_w) =$
$\quad C_{p,m} m_m (T_m - T_f)$

$C_{p,m} = \dfrac{C_{p,w} m_w (T_f - T_w)}{m_m (T_m - T_f)}$

$C_{p,m} =$

$\dfrac{(4186 \text{ J/kg} \cdot {}^\circ\text{C})(0.150 \text{ kg})(29.0 {}^\circ\text{C} - 25.0 \, {}^\circ\text{C})}{(7.50 \times 10^{-2} \text{ kg})(93.0 {}^\circ\text{C} - 29.0 {}^\circ\text{C})}$

$C_{p, m} =$

$\dfrac{(4186 \text{ J/kg} \cdot {}^\circ\text{C})(0.150 \text{ kg})(4.0 {}^\circ\text{C})}{(7.50 \times 10^{-2} \text{ kg})(64.0 {}^\circ\text{C})} =$

$\boxed{5.2 \times 10^2 \text{ J/kg} \cdot {}^\circ\text{C}}$

10 Thermodynamics

RELATIONSHIPS BETWEEN HEAT AND WORK

1. c
2. b
3. a
4. c

Given

$P = 1.5 \times 10^5 \text{ Pa}$

$\Delta V = -30 \times 10^{-3} \text{ m}^3$

Solution

$W = P\Delta V =$
$\quad (1.5 \times 10^5 \text{ Pa})(-3.0 \times 10^{-3} \text{ m}^3) =$
$\quad -450 \text{ J}$

5. d
6. b
7. c
8. a

9. Energy that is added or removed from the system during an isothermal process causes a slight change in the internal energy of the system by increasing or decreasing the kinetic energy of the particles in the system. However, this energy is almost immediately transferred from or to the system, so that the internal energy, and thus the temperature of the system does not effectively change.

10. -1.8×10^3 J

Given

$P = 3.0 \times 10^5$ Pa

$V_i = 1.55 \times 10^{-2}$ m^3

$V_f = 9.5 \times 10^{-3}$ m^3

Solution

$W = P\Delta V = P(V_f - V_i)$

$W = (3.0 \times 10^5 \text{ Pa})$
$\quad (9.5 \times 10^{-3} \text{ m}^3 - 1.55 \times 10^{-2} \text{ m}^3)$

$W = (3.0 \times 10^5 \text{ Pa})(-6.0 \times 10^{-3} \text{ m}^3)$

$\quad = \boxed{-1.8 \times 10^3 \text{ J}}$

10 Thermodynamics

THE FIRST LAW OF THERMODYNAMICS

1. b
2. c
3. a
4. d
5. b
6. d

Given

$Q_{added} = 2750$ J

$Q_{removed} = 1550$ J

$W = 850$ J

Solution

$\Delta U = Q - W = (2750 \text{ J} - 1550 \text{ J}) - 850 \text{ J} = 350 \text{ J}$

7. c
8. a
9. In a cyclic process, the temperature of the system does change, unlike an isothermal process. However, once the cycle is complete, the system is at the same temperature as when it started the cycle, so the overall effect is as if there was no change in temperature. Like an isothermal process, the overall energy transferred as heat equals the overall work performed.

10. 1.36×10^5 J

Given

$Q_{added} = 6.60 \times 10^5$ J

$Q_{removed} = 4.82 \times 10^5$ J

$\Delta U = 4.2 \times 10^4$ J

Solution

$\Delta U = Q - W$

$Q = Q_{added} - Q_{removed}$

$W = Q_{added} - Q_{removed} - \Delta U =$
$\quad 6.60 \times 10^5 \text{ J} - 4.82 \times 10^5 \text{ J} -$
$\quad 4.2 \times 10^4 \text{ J}$

$W = \boxed{1.36 \times 10^5 \text{ J}}$

10 Thermodynamics

THE SECOND LAW OF THERMODYNAMICS

1. b
2. a
3. d
4. d

Given

$Q_c = \dfrac{2}{3}Q_h$

Solution

$eff = 1 - \dfrac{Q_c}{Q_h} = 1 - \dfrac{\frac{2}{3}Q_h}{Q_h} = 1 - \dfrac{2}{3} = \dfrac{1}{3}$

5. c

Given

$Q_h = 2.5 \times 10^4$ J

$W_{net} = 7.0 \times 10^3$ J

Solution

$eff = \dfrac{W_{net}}{Q_h} = \dfrac{7.0 \times 10^3 \text{ J}}{2.5 \times 10^4 \text{ J}} = 0.28$

6. b
7. a
8. a
9. According to the second law of thermodynamics, energy (Q_c) must be given up as heat to a lower-temperature body from a heat engine. Because Q_c is always greater than zero, the ratio of energy transferred from the engine as heat to the energy transferred to the engine as heat (Q_c/Q_h) is greater than zero. For an engine to be 100 percent efficient, this ratio must be zero. Therefore, the second law of thermodynamics predicts that no heat engine can be 100 percent efficient.

10. 0.25

Given

$Q_h = 7.6 \times 10^5$ J

$Q_c = 5.7 \times 10^5$ J

Solution

$$eff = 1 - \frac{Q_c}{Q_h}$$

$$eff = 1 - \frac{5.7 \times 10^5 \text{ J}}{7.6 \times 10^5 \text{ J}} = 1 - 0.75 =$$

$\boxed{0.25}$

11 Vibrations and Waves

SIMPLE HARMONIC MOTION

1. b   **5.** a

2. a   **6.** c

3. d   **7.** a

4. d   **8.** d

9. The system eventually stops oscillating and comes to rest.

10. 43 N/m

Given

$m = 370$ g $= 0.37$ kg

$x = 8.5$ cm $= 0.085$ m

Solution

$$k = \frac{-mg}{x}$$

$$k = \frac{-(0.37 \text{ kg})(9.81 \text{ m/s}^2)}{-0.085 \text{ m}}$$

$$k = \boxed{43 \text{ N/m}}$$

11 Vibrations and Waves

MEASURING SIMPLE HARMONIC MOTION

1. d   **5.** d

2. a   **6.** a

3. c   **7.** b

4. c   **8.** c

9. Answers will vary. Sample answer: The period (and frequency) of pendulums depends only on the length, provided that both pendulums are subject to the same acceleration of gravity. Restoring force does increase with increasing mass, but the acceleration stays the same.

10. $T = 2.19$ s; $f = 0.457$ Hz

Given

$L = 1.20$ m

$a_g = 9.81$ m/s^2

Solution

For a pendulum,

$$T = 2\pi\sqrt{\frac{L}{a_g}} = 2\pi\sqrt{\frac{1.20 \text{ m}}{9.81 \text{ m/s}^2}} = 2.19 \text{ s}$$

$$f = \frac{1}{T} = \frac{1}{2.19 \text{ s}} = \boxed{0.457 \text{ Hz}}$$

11 Vibrations and Waves

PROPERTIES OF WAVES

1. a   **5.** b

2. d   **6.** b

3. c   **7.** d

4. b   **8.** a

9. The frequency does not change because it is determined by the source of the wave motion. Therefore, the wavelength increases because the velocity increases.

10. 3.14 m

Given

$f = 95.5$ MHz $= 9.55 \times 10^7$ Hz

$v = 3.00 \times 10^8$ m/s

Solution

$$\lambda = \frac{v}{f} = \frac{3.00 \times 10^8 \text{ m/s}}{9.55 \times 10^7 \text{ Hz}} = \boxed{3.14 \text{ m}}$$

11 Vibrations and Waves

WAVE INTERACTIONS

1. a   **5.** b

2. d   **6.** d

3. c   **7.** c

4. d   **8.** a

9. If the displacements are in the same direction (both are crests or both are troughs), constructive interference will occur. If the displacements are in opposite directions (one is a crest and the other is a trough), destructive interference will occur.

10. The boundary must be fixed; i.e. the rope must be firmly attached to the pole, unable to slide up and down.

12 Sound

SOUND WAVES

1. b   **5.** c

2. a   **6.** d

3. c   **7.** a

4. a   **8.** b

9. As the object moves in one direction, it pushes against the air in that direction, forming a compression by squeezing the molecules closer together. As the object moves in the opposite direction, the air molecules spread apart, forming a rarefaction. As the object continues to vibrate, a series of compressions and rarefactions travel through the air, creating a sound wave.

10. Wave 1. The waves are traveling at the same speed if they are in the same medium. Therefore, the wave with the shorter wavelength (Wave 1) is created by the higher frequency.

12 Sound

SOUND INTENSITY AND RESONANCE

1. b   5. b
2. d   6. c
3. c   7. a
4. c   8. d

9. Damage to the ears can result from prolonged exposure to sounds that are not loud enough to cause immediate damage.

10. 19 W

Given

Intensity $= 4.6 \times 10^{-3}$ W/m^2

$r = 18$ m

Solution

intensity $= \dfrac{P}{4\pi r^2}$; $P = 4\pi r^2$ (intensity)

$P = 4\pi(18 \text{ m})^2(4.6 \times 10^{-3} \text{ W/m}^2) =$

$\boxed{19 \text{ W}}$

12 Sound

HARMONICS

1. d
2. d
3. b
4. c
5. c
6. a

Given

$v = 488$ m/s

$f_1 = 1250$ Hz

Solution

$f_n = n\dfrac{v}{2L}$; $2Lf_n = nv$; $L = \dfrac{nv}{2f_n}$

For the fundamental frequency, $n = 1$, so

$L = \dfrac{v}{2f_1} = \dfrac{488 \text{ m/s}}{2(1250 \text{ Hz})} = 0.195 \text{ m} =$

19.5 cm

7. d
8. c
9. The sound waves of the two notes interfere constructively and then destructively resulting in alternating loud and soft moments. The number of these louder-softer combinations that occur per second is equal to the difference in frequencies of the two notes.

10. 17.2 Hz

Given

$L = 10.0$ m

$v = 344$ m/s

Solution

For a pipe that is open at both ends,

$f_n = n\dfrac{v}{2L}$

At the fundamental frequency, $n = 1$, so

$f_1 = 1\left(\dfrac{v}{2L}\right) = \dfrac{344 \text{ m/s}}{2(10.0 \text{ m})} = \boxed{17.2 \text{ Hz}}$

13 Light and Reflection

CHARACTERISTICS OF LIGHT

1. d   5. b
2. a   6. a
3. c   7. c
4. a   8. b

9. Answers may vary. Sample answer: there is no sharp division between one kind of wave and the next.

10. 4.6×10^{11} Hz

Given

$\lambda = 650$ μm $= 650 \times 10^{-6}$ m $=$
6.5×10^{-4} m

$c = 3.00 \times 10^8$ m/s

Solution

Rearrange the wave speed equation and substitute values.

$c = f\lambda$

$f = c/\lambda = \dfrac{(3.00 \times 10^8 \text{ m/s})}{(6.5 \times 10^{-4} \text{ m})} =$

$\boxed{4.6 \times 10^{11} \text{ Hz}}$

13 Light and Reflection

FLAT MIRRORS

1. b	**5.** c
2. d	**6.** d
3. a	**7.** b
4. b	**8.** b

9. Answers may vary. Sample answer: Virtual; the rays that form the image appear to come from a point behind the mirror.

10.

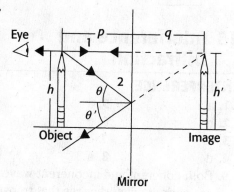

13 Light and Reflection

CURVED MIRRORS

1. a	**5.** b
2. b	**6.** c
3. c	**7.** d
4. b	**8.** a

9. Answers may vary. Sample answer: A spherical mirror is a portion of a spherical shell. In contrast, a parabolic mirror is made from segments of a reflecting paraboloid. With a parabolic mirror, all rays parallel to the principal axis converge at the focal point regardless of where on the mirror's surface the rays reflect. Thus, a real image forms without spherical aberration.

10. 33.3 cm

Given

$f = +20.0$ cm

$p = +50.0$ cm

The mirror is concave, so f is positive. The object is in front of the mirror, so p is positive.

Solution

Rearrange the equation to isolate the image distance, and calculate.

$1/q = 1/f - 1/p$

$1/q = 1/20.0 \text{ cm} - 1/50.0 \text{ cm} =$

$0.0500/1 \text{ cm} - 0.0200/1 \text{ cm} =$

$0.0300/1 \text{ cm}$

$q = \boxed{33.3 \text{ cm}}$

13 Light and Reflection

COLOR AND POLARIZATION

1. c	**5.** b
2. a	**6.** a
3. d	**7.** c
4. b	**8.** d

9. Answers may vary. Sample answer: In the correct proportions, a mixture of the three primary pigments produces a black mixture because all colors are subtracted or absorbed from white light.

10. Answers may vary. Sample answer: By rotating a polarizing substance in the beam of light. If the light intensity changes as the polarizing substance is rotated and eventually no light can pass through, the beam of light is polarized.

14 Refraction

REFRACTION

1. d	**5.** a
2. c	**6.** b
3. a	**7.** c
4. b	**8.** d

9. Answers may vary. Sample answer: As wave fronts enter a transparent medium, they slow down, but the wave fronts that have not yet reached the surface of the medium continue to move at the same speed. During this time the slower wave fronts travel a smaller distance than do the original wave fronts, so the entire plane wave changes directions.

10. 31.6°

Given

$n_i = 1.333$

$n_r = 1.458$

$\theta_i = 35.0°$

Solution

Rearrange the Snell's law, $n_i \sin \theta_i = n_r \sin \theta_r$, and solve for θ_r.

$$\theta_r = \sin^{-1}\left[\frac{n_i}{n_r}(\sin \theta_i)\right] =$$

$$\sin^{-1}\left[\frac{1.333}{1.458}(\sin 35.0°)\right] = \boxed{31.6°}$$

14 Refraction

THIN LENSES

1. b	**5.** c
2. b	**6.** b
3. a	**7.** c
4. d	**8.** c

9. Answers may vary. Sample answer: The image formed by the first lens is treated as the object for the second lens.

10. 1.00×10^{-2} cm

Given

$p = 25.0$ cm

$f = 20.0$ cm

Solution

Rearrange the thin lens equation, $\frac{1}{p} + \frac{1}{q} = \frac{1}{f}$, and solve for q.

$$\frac{1}{q} = \frac{1}{f} - \frac{1}{p} = \frac{1}{20.0 \text{ cm}} - \frac{1}{25.0 \text{ cm}} =$$

$$\frac{0.0500}{1 \text{ cm}} - \frac{0.0400}{1 \text{ cm}} = \frac{0.0100}{1 \text{ cm}}$$

$$q = \boxed{1.00 \times 10^{-2} \text{ cm}}$$

14 Refraction

OPTICAL PHENOMENA

1. c	**5.** d
2. d	**6.** a
3. b	**7.** a
4. b	**8.** d

9. Answers may vary. Sample answer: When an observer views a raindrop high in the sky, the red light reaches the observer, but the violet light, like the other spectral colors, passes over the observer because it deviates from the path of the white light more than the red light does.

10. 47.29°

Given

$n_i = 1.361$

$n_r = 1.000$

Solution

Use the critical angle equation, $\sin \theta_c = n_r/n_i$, and solve for c.

$$\theta_c = \sin^{-1}\left(\frac{n_r}{n_i}\right) = \sin^{-1}\left(\frac{1.000}{1.361}\right) =$$

$$\boxed{47.29°}$$

15 Interference and Diffraction

INTERFERENCE

1. a	**5.** d
2. b	**6.** c
3. a	**7.** c
4. d	**8.** a

9. Both coherent and incoherent waves are periodic disturbances that transfer energy. Coherent waves have wavelengths that are equal and travel in phase. Incoherent waves usually do not have equal wavelengths and do not travel in phase.

10. 610 nm

Given

$d = 2.1 \times 10^{-3}$ mm

$\theta = 17.0°$

$m = 1$

Solution

$$d \sin \theta = m\lambda$$

$$\lambda = \frac{d \sin \theta}{m} =$$

$$\frac{(2.1 \times 10^{-3} \text{ mm})(\sin 17.0°)}{1} =$$

$$6.1 \times 10^{-4} \text{ mm} = \boxed{6.1 \times 10^2 \text{ nm}}$$

15 Interference and Diffraction

DIFFRACTION

1. c	**5.** a
2. b	**6.** c
3. d	**7.** b
4. a	**8.** d

9. Even though the waves are of the same wavelength and are in phase, wavelets that travel different distance to the screen, such as one from the center and one from the edge of the slit, are out of phase and therefore can undergo destructive interference.

10. $17°$

Given

$d = 4.0 \times 10^{-4}$ cm $= 4.0 \times 10^{-6}$ m

$\lambda = 600$ nm $= 6.0 \times 10^{-7}$ m

$m = 2$

Solution

$d\sin\theta = m\lambda$

$\theta = \sin^{-1}\left(\dfrac{m\lambda}{d}\right) =$

$\sin^{-1}\left(\dfrac{2(6.0 \ 10^{-7} \text{ m})}{4.0 \times 10^{-6} \text{ m}}\right) = \boxed{17°}$

15 Interference and Diffraction

LASERS

1. c   **5.** b
2. a   **6.** b
3. d   **7.** a
4. a   **8.** a

9. The term laser represents the words "light amplification by the stimulated emission of radiation."

10. An acceptable answer should include a wave that has the same wavelength and phase as the original wavelength but may vary in amplitude.

16 Electric Forces and Fields

ELECTRIC CHARGE

1. b   **5.** a
2. a   **6.** d
3. b   **7.** c
4. c   **8.** b

9. The electrons in a conductor are free to move from place to place, whereas the electrons in insulators cannot move freely.

10. induction

16 Electric Forces and Fields

ELECTRIC FORCE

1. c   **5.** d
2. d   **6.** d
3. d   **7.** a
4. b   **8.** b
9. vector

10. -5.0×10^{-2} N

Given

$q_1 = 4.0 \times 10^{-6}$ C

$q_2 = -8.0 \times 10^{-6}$ C

$r = 2.4 \times 10^{-2}$ m

$k_C = 8.99 \times 10^9$ N•m²/C²

Solution

$F_{electric} = k_C \dfrac{q_1 q_2}{r^2} =$

$(8.99 \times 10^9 \text{ N m}^2/\text{C}^2) \times$

$\left(\dfrac{(4.0 \times 10^{-6} \text{ C})(-8.0 \times 10^{-6} \text{ C})}{(2.4 \times 10^{-2} \text{ m})^2}\right)$

$F_{electric} = \boxed{-5.0 \times 10^2 \text{ N}}$

16 Electric Forces and Fields

THE ELECTRIC FIELD

1. c   **5.** d
2. c   **6.** d
3. c   **7.** b
4. a   **8.** b

9. The sign of the charge on the left is negative. The diagram shows the field lines curving toward the positive charge on the right. They curve that way only when the charge on the left is an opposite charge.

10. 1.39×10^{-2} m

Given

$E = 2250$ N/C

$q = 4.82 \times 10^{-11}$ C

$k_C = 8.99 \times 10^9$ N•m²/C²

Solution

$E = k_C \dfrac{q}{r^2}$

Rearrange to solve for r.

$r = \sqrt{k_C \dfrac{q}{E}} =$

$\sqrt{(8.99 \times 10^9 \text{ N•m}^2/\text{C}^2)\left(\dfrac{(4.82 \times 10^{-11} \text{ C})}{2250 \text{ N/C}}\right)}$

$r = \boxed{1.39 \times 10^{-2} \text{ m}}$

17 Electrical Energy and Current

ELECTRIC POTENTIAL

1. a   **5.** d
2. c   **6.** b
3. c   **7.** b
4. c   **8.** d

9. The chemical energy moves negative charges from the positive terminal to the negative terminal. This increases the electrical potential energy of the charges.

10. 8.6×10^4 V

Given
$q = 7.6 \times 10^{-9}$ C
$r = 7.9 \times 10^{-4}$ m

Solution
Use the equation for the potential difference between a point at infinity and a point near a charge.

$$\Delta V = k_C \frac{q}{r} = (8.99 \times 10^9 \text{ N m}^2/\text{C}^2) \times$$

$$\left(\frac{7.6 \times 10^{-9} \text{ C}}{7.9 \times 10^{-4} \text{ m}} \right) = \boxed{8.6 \times 10^4 \text{ V}}$$

17 Electrical Energy and Current

CAPACITANCE

1. b   **5.** a
2. c   **6.** d
3. c   **7.** d
4. c   **8.** b

9. Charges move from one plate of the capacitor through the connecting wire to the other plate until the two plates are no longer charged.

10. 4.4×10^2 J

Given
$C = 2.2 \times 10^{-3}$ F
$\Delta V = 450$ V

Solution
$PE_{electric} = \frac{1}{2} C(\Delta V)^2 =$

$\frac{1}{2} (2.2 \times 10^{-3} \text{ F})(450)^2 =$

$\frac{1}{2} (2.2 \times 10^{-3} \text{ F})(2.0 \times 10^5) =$

$$\boxed{4.4 \times 10^2 \text{ J}}$$

17 Electrical Energy and Current

CURRENT AND RESISTANCE

1. c   **5.** c
2. a   **6.** c
3. b   **7.** b
4. c   **8.** c

9. The electric field that provides the energy for the light moves at nearly the speed of light.

10. 1.28 s

Given
$I = 295$ A
$q = 377$ C

Solution
$$\Delta t = \frac{q}{I} = \frac{377 \text{ C}}{295 \text{ A}} = \boxed{1.28 \text{ s}}$$

17 Electrical Energy and Current

ELECTRIC POWER

1. b   **5.** b
2. b   **6.** a
3. c   **7.** c
4. d   **8.** d

9. With alternating current, electrons vibrate back and forth instead of flowing in one direction. The energy of this vibration does useful work.

10. $3.2

Given
$\Delta V = 117$ V
$I = 8.7$ A
$\Delta t = 28$ h
Energy Cost = $0.115 per kW•h

Solution
$P = \Delta V \times I = 8.7 \text{ A} \times 117 \text{ V} =$
1.0×10^3 W

$kW = P \times \dfrac{1 \text{ kW}}{1000 \text{ W}} =$

$(1.0 \times 10^3 \text{ W}) \times \dfrac{1 \text{ kW}}{1000 \text{ W}} = 1 \text{ kW}$

$\dfrac{1 \text{ h}}{\text{day}} \times 28 \text{ days} = 28 \text{ h}$

$1 \text{ kW} \times 28 \text{ h} = 28 \text{ kW•h}$

$1 \text{ kWh} \times \textit{Energy Cost} =$

$28 \text{ kW•h} \times \dfrac{\$0.115}{\text{kW•h}} = \boxed{\$3.2}$

18 Circuits and Circuit Elements

SCHEMATIC DIAGRAMS AND CIRCUITS

1. b
2. c
3. c
4. d

5. c
6. b
7. c
8. b

9. Student answers will vary, but the depiction of the named components should agree with the schematic symbols in Table 1 of the text. One possible schematic is shown.

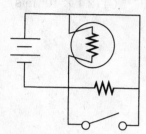

10. each battery : 1.2 V, radio : 12.0 V

Given

batteries = 10
$\Delta V_{total} = 12.0$ V

Solution

$$\Delta V_{battery} = \frac{\Delta V_{total}}{batteries} = \frac{12.0 \text{ V}}{10} = 1.2 \text{ V}$$

$$\Delta V_{radio} = \Delta V_{total} = \boxed{12.0 \text{ V}}$$

18 Circuits and Circuit Elements

RESISTORS IN SERIES OR IN PARALLEL

1. b
2. a
3. b
4. c

5. b
6. c
7. c
8. b

9. Divide the resistance of one resistor by the number of resistors in the circuit to get the equivalent resistance.

10. 214 Ω

Given

$R_a = 38.7 \ \Omega$
$R_b = 89.5 \ \Omega$
$I_a = 0.155$ A
$I_{tot} = 0.250$ A

Solution

Find the voltage across the circuit using the current and resistance of R_a.

$\Delta V = I_a R_a = (0.155 \text{ A}) \times (38.7 \ \Omega) =$
6.00 V

Find the current in R_b.

$$I_b = \frac{\Delta V}{R_b} = \frac{6.00 \text{ V}}{89.5 \ \Omega} = 0.0670 \text{ A}$$

Find the current through R_c.

$I_c = I_{tot} - (I_a + I_b) = 0.250 \text{ A} -$
$(0.155 \text{ A} + 0.0670 \text{ A}) = 0.0280 \text{ A}$

Calculate the resistance for R_c.

$$R_c = \frac{\Delta V}{I_c} = \frac{6.00 \text{ V}}{0.0280 \text{ A}} = \boxed{214 \ \Omega}$$

18 Circuits and Circuit Elements

COMPLEX RESISTOR COMBINATIONS

1. b
2. c
3. c
4. b

5. c
6. b
7. a
8. d

9. Student answers may vary. However, all answers should include reducing the circuit to smaller groups of series or parallel combinations, calculating the equivalent resistance of the groups, and finally the resistance of the entire circuit. There is no rule as to where to start or the order of the steps to take.

10. 4.5 V

Given

$R_a = 5.6 \ \Omega$
$R_b = 8.2 \ \Omega$
$R_c = 3.3 \ \Omega$
$R_d = 7.5 \ \Omega$
$R_e = 4.7 \ \Omega$
$I_c = 0.44$ A

Solution

Find the equivalent resistance of R_a and R_b.

$$\frac{1}{R_x} = \frac{1}{5.6 \ \Omega} + \frac{1}{8.2 \ \Omega} = \frac{0.18}{1 \ \Omega} + \frac{0.12}{1 \ \Omega} =$$

$$\frac{0.30}{1 \ \Omega}; R_x = \frac{1 \ \Omega}{0.30} = 3.3 \ \Omega$$

Find the equivalent resistance of R_c and R_d.

$$\frac{1}{R_y} = \frac{1}{3.3 \ \Omega} + \frac{1}{7.5 \ \Omega} = \frac{0.30}{1 \ \Omega} +$$

$$\frac{0.13}{1 \ \Omega} = \frac{0.43}{1 \ \Omega}; R_y = \frac{1 \ \Omega}{0.43} = 2.3 \ \Omega$$

Find the equivalent resistance of the circuit.

$$R_{eq} = R_x + R_y + R_e = 3.3 \ \Omega + 2.3 \ \Omega + 4.7 \ \Omega = 10.3 \ \Omega$$

Use Ohm's law to find the battery voltage.

$$\Delta V = I_c \times R_{eq} : 0.44 \ A \times 10.3 \ \Omega = $$

$$\boxed{4.5 \ V}$$

19 Magnetism

MAGNETS AND MAGNETIC FIELDS

1. b   5. d
2. d   6. a
3. c   7. a
4. b   8. c

9. (a) attract; (b) attract; (c) repel; (d) attract

10. Lines should be drawn from the magnet's north pole to its south pole. A greater number of lines should be drawn leaving and entering the magnet's north and south pole, respectively.

19 Magnetism

MAGNETISM FROM ELECTRICITY

1. a   5. d
2. d   6. c
3. b   7. d
4. b   8. c

9. Answers may vary. Sample answer: A long, helically wound coil of insulated wire; electromagnet; A more powerful magnet is created.

10. External lines should extend out from the figure's right end and enter the figure's left end. Internal lines should be close together and nearly parallel. The right end of the figure should be labeled N while the left end is labeled S.

19 Magnetism

MAGNETIC FORCE

1. c
2. c
3. d

Given
$F_{magnetic} = 3.8 \times 10^{-13} \ N$
$q = 1.60 \times 10^{-19} \ C$
$v = 2.4 \times 10^6 \ m/s$

Solution
$$B = \frac{F_{magnetic}}{qv}$$
$$B = \frac{(3.8 \times 10^{-13} \ N)}{(1.60 \times 10^{-19} \ C)(2.4 \times 10^6 \ m/s)}$$

$$= 0.99 \ T$$

4. c
5. b
6. d
7. d
8. a

Given
$B = 6.0 \times 10^{-4} \ T$
$I = 25 \ A$
$\ell = 10.0 \ m$

Solution
$F_{magnetic} = BI\ell$
$F_{magnetic} = (6.0 \times 10^{-4} \ T)(25 \ A)$
$\quad (10.0 \ m) = 1.5 \times 10^{-1} \ N$

9. Answers may vary. Sample answer: The electric particles moving through the wire are also moving through the magnetic field. The resultant force on the wire is the sum of the individual magnetic forces on the charged particles.

10. $1.5 \times 10^6 \ m/s$

Given
$F_{magnetic} = 8.2 \times 10^{-16} \ N$
$q = 1.60 \times 10^{-19} \ C$
$B = 3.5 \times 10^{-3} \ T$

Solution
$$B = \frac{F_{magnetic}}{qv}$$

Rearrange the equation to solve for v:

$$v = \frac{F_{magnetic}}{qB}$$
$$v = \frac{(8.2 \times 10^{-16} \ N)}{(1.60 \times 10^{-19} \ C)(3.5 \times 10^{-3} \ T)}$$

$$= \boxed{1.5 \times 10^6 \ m/s}$$

20 Electromagnetic Induction

ELECTRICITY FROM MAGNETISM

1. d	**5.** a
2. b	**6.** b
3. c	**7.** c
4. d	**8.** a

9. Answers may vary. Sample answer: A moving charge can be deflected by a magnetic field and as long as the motion of the charge is not parallel to the magnetic field the charge experiences a magnetic force. The effect of this force is equivalent to replacing the moving segment of wire and the magnetic field with a battery or other electrical power supply.

10. -180 V

Given

$N = 150$ turns
$\Delta t = 1.1$ s
$B_i = 0.00$ T
$B_f = 0.95$ T
$A = 1.4$ m^2
$\theta = 0.0°$

Solution

Substitute values into Faraday's law of magnetic induction.

$$\text{emf} = -\frac{\Delta \Phi_M}{\Delta t} = -\frac{N\Delta[AB\cos\theta]}{\Delta t} =$$

$$-NA\cos\theta\left[\frac{\Delta B}{\Delta t}\right] =$$

$$-NA\cos\theta\left[\frac{(B_f - B_i)}{\Delta t}\right]$$

$$= -(150)\,(1.4\ \text{m}^2)\,(\cos 0.0°)$$

$$\left[\frac{(0.95\ \text{T} - 0.00\ \text{T})}{(1.1\text{s})}\right] = \boxed{-180\ \text{V}}$$

20 Electromagnetic Induction

GENERATORS, MOTORS, AND MUTUAL INDUCTANCE

1. b	**5.** d
2. c	**6.** c
3. b	**7.** b
4. a	**8.** a

9. Falling water or steam produced by heating water turns the blades of a turbine.

10. Answers may vary. Sample answer: Mutual inductance is the process by which a changing magnetic field in the primary circuit's coil induces a current in the secondary circuit's coil and vice versa. Transformer.

20 Electromagnetic Induction

AC CIRCUITS AND TRANSFORMERS

1. c	**5.** d
2. a	**6.** b
3. b	**7.** c
4. d	**8.** a

9. Power companies use a high emf and a low current. Since the power lost is equal to I^2R, reducing the current by a factor of 10, reduces the power lost by a factor of 100. Transformer.

10. $\Delta V_{rms} = 160$ V
$I_{rms} = 1.5$ A
$I_{max} = 2.1$ A

Given

$\Delta V_{max} = 230$ V
$R = 110\Omega$

Solution

$\Delta V_{rms} = 0.707\ \Delta V_{max} = (0.707)(230\ \text{V})$

$= \boxed{160\ \text{V}}$

$I_{rms} = \Delta V_{rms}/R = 160\text{V}/\ 110\Omega$

$= \boxed{1.5\ \text{A}}$

$I_{max} = I_{rms}/0.707 = 1.5\ \text{A}/0.707$

$= \boxed{2.1\ \text{A}}$

20 Electromagnetic Induction

ELECTROMAGNETIC WAVES

1. a	**5.** a
2. d	**6.** d
3. b	**7.** b
4. b	**8.** c

9. Answers may vary. Sample answer: The electromagnetic spectrum is a graphic way to represent different types of electromagnetic waves and illustrate the relationship of these waves in terms of wavelength, frequency, and energy.

10. Answers may vary. Sample answer: Microwaves have long wavelengths, low frequency, and relatively low energy. Microwaves are used to cook food, transmit telephone, computer and satellite data, and radar.

21 Atomic Physics

QUANTIZATION OF ENERGY

1. d
2. b
3. b
4. a

Given

$f = 5.45 \times 10^{14}$ Hz
$h = 6.63 \times 10^{-34}$ J•s

Solution

$E = hf =$
$(6.63 \times 10^{-34} \text{ J•s})(5.45 \times 10^{14} \text{ Hz})$
$= 3.61 \times 10^{-19}$ J

5. c

Given

$E = 1.3 \times 10^{-19}$ J
$h = 6.63 \times 10^{-34}$ J•s

Solution

$f = \dfrac{E}{h} = \dfrac{1.3 \times 10^{-19} \text{ J}}{6.63 \times 10^{-34} \text{ J•s}} =$
2.0×10^{14} Hz

6. d
7. d

Given

$hf = 3.7$ eV
$hf_t = 3.5$ eV

Solution

$KE_{max} = hf - hf_t = 3.7 \text{ eV} - 3.5 \text{ eV}$
$= 0.2$ eV

8. c
9. Classical electromagnetic theory predicted that, as the wavelength of light approached zero, the amount of energy emitted by a blackbody would become infinite. By limiting the frequencies that a blackbody could radiate to fixed integral increments, Planck was able to modify the classical theory so that radiated energy reached a maximum at a particular frequency and then decreased toward zero, as was indicated by experimental results.

10. 0.3 eV

Given

$f = 6.6 \times 10^{14}$ Hz
$hf_t = 2.4$ eV
$h = 6.63 \times 10^{-34}$ J•s
$1 \text{ eV} = 1.6 \times 10^{-19}$ J

Solution

$KE_{max} = hf - hf_t$
$KE_{max} =$
$(6.63 \times 10^{-34} \text{ J•s})(6.6 \times 10^{14} \text{ Hz}) \times$
$\dfrac{1 \text{eV}}{1.60 \times 10^{-19} \text{ J}} - 2.4$ eV
$KE_{max} = 2.7 \text{ eV} - 2.4 \text{ eV} = \boxed{0.3 \text{ eV}}$

21 Atomic Physics

MODELS OF THE ATOM

1. b
2. b
3. d
4. c
5. c
6. a
7. b
8. c

9. The Rutherford model of the atom had a compact, massive nucleus that contained the positive charge of the atom. The electrons orbited this nucleus in a manner similar to planets orbiting the sun. This model correctly accounted for the nucleus, but according to classical electromagnetic theory, the orbiting electrons would radiate energy continuously, and so could not maintain stable orbits. Also, Rutherford's model provided no explanation for emission or absorption spectra.

10. In the Bohr model of the atom, electrons move in orbits around the nucleus, much like planets orbit the sun. However, only specific orbits are stable and can thus be occupied by the electrons. When electrons are in these orbits, they do not emit radiation, as predicted by classical electromagnetic theory. Radiation is emitted when electrons move from higher-energy levels to lower-energy levels, which accounts for the individual lines of emission spectra. However, Bohr's model could not explain why the stable orbits existed, or why radiation was not emitted when the electrons were in these orbits. The Bohr model also did not predict the spectra for atoms with more than one electron in their outermost energy levels.

21 Atomic Physics

QUANTUM MECHANICS

1. a

2. b

3. b

4. c

Given

$\lambda = 6.63 \times 10^{-9}$ m

h $= 6.63 \times 10^{-34}$ J•s

Solution

$$p = \frac{h}{\lambda} = \frac{6.63 \times 10^{-34} \text{ J•s}}{6.63 \times 10^{-9} \text{ m}} =$$

1.00×10^{-25} kg•m/s

5. d

6. c

7. c

8. d

9. The idea of photons provided an explanation for both blackbody radiation and the photoelectric effect by using the concept of wave-particle duality. Light had both the characteristics of waves and particles, and under certain conditions would exhibit primarily one characteristic or the other. The idea of matter waves extended this idea to matter, which had till that time been thought of exclusively as particles. By giving matter characteristics of waves, wave-particle duality was extended to describe both light and matter.

10. 14.1 pm

Given

$m = 1.88 \times 10^{-28}$ kg

$v = 2.50 \times 10^5$ m/s

h $= 6.63 \times 10^{-34}$ J•s

Solution

$$\lambda = \frac{h}{p} = \frac{h}{mv}$$

$$\lambda = \frac{6.63 \times 10^{-34} \text{ J•s}}{(1.88 \times 10^{-28} \text{ kg})(2.50 \times 10^5 \text{ m/s})}$$

$= 1.41 \times 10^{-11}$ m $= \boxed{14.1 \text{ pm}}$

22 Subatomic Physics

THE NUCLEUS

1. b   **4.** a

2. d   **5.** c

3. b   **6.** c

7. b

Given

For $^{52}_{24}$Cr:

$Z = 24$

$A = 52$

atomic mass of Cr-52 $= 51.940\ 511$ u

atomic mass of H $= 1.007\ 825$ u

atomic mass of $m_n = 1.008\ 665$ u

Solution

$N = A - Z = 52 - 24 = 28$

$\Delta m = Z$(atomic mass of $^1_1$H) $+$
Nm_n − atomic mass

$= 24(1.007\ 825$ u$) + 28(1.008\ 665$ u$) -$
$51.940\ 511$ u

$= 24.187\ 800$ u $+ 28.242\ 620$ u $-$
$51.940\ 511$ u $= \boxed{0.489\ 909 \text{ u}}$

8. d

Given

$\Delta m = 0.466\ 769$ u

$c^2 = 931.49$ MeV/u

Solution

$E_{bind} = \Delta mc^2 (0.466\ 769$ u$)$
$(931.49$ MeV/u$) = \boxed{434.79 \text{ MeV}}$

9. The more-massive nuclei have more protons in their nucleus, and so have greater repulsive forces within them. More neutrons are needed to provide more of the strong force that holds nucleons together without causing any more repulsion.

10. 167.41 MeV

Given

For $^{21}_{10}$Ne:

$Z = 10$

$A = 21$

atomic mass of Ne-21 $= 20.993\ 841$ u

Solution

$N = A - 2 = 21 - 10 = 11$

$\Delta m = Z$(atomic mass of $^1_1$H) $+ Nm_n -$
atomic mass

$= 10(1.007\ 825$ u$) + 11(1.008\ 665$ u$) -$
$20.993\ 841$ u

$= 10.078\ 250$ u $+ 11.095\ 315$ u $-$
$20.993\ 841$ u

$= 0.179\ 724$ u

$E_{bind} = (0.179\ 724$ u$)(931.49$ MeV/u$) =$
$\boxed{167.41 \text{ MeV}}$

22 Subatomic Physics

NUCLEAR DECAY

1. b
2. a
3. a
4. c
5. b

Solution

Mass number of unknown =
    $232 - 4 = 228$
Atomic number of unknown =
    $90 - 2 = 88$
From the periodic table, the nucleus with an atomic number of 88 is Ra, so $^{228}_{88}$Ra is the unknown decay product.

6. a

Solution

Mass number of unknown =
    $14 - 0 = 14$
Atomic number of unknown =
    $6 - (-1) = 7$
From the periodic table, the nucleus with an atomic number of 7 is N, so $^{14}_{7}$N is the unknown decay product.

7. b
8. d

Solution

In 3 half-lives, the sample will be diminished by $\frac{1}{2^3} = \frac{1}{8}$, so $\frac{24 \text{ g}}{8} = \boxed{3 \text{ g}}$.

9. In a nuclear decay process, charge and the number of nucleons must be conserved. To conserve charge, the total of the atomic numbers on the left side of the equation must be the same as the total of the atomic numbers on the right. To conserve the nucleon number, the total of the mass numbers on the left must be the same as the total of the mass numbers on the right.

10. 6.9×10^{16} Bq

Given

$T_{1/2} = 1.6$ s
$N = 1.6 \times 10^{17}$

Solution

$$T_{1/2} = \frac{0.693}{\lambda}$$

$$\lambda = \frac{0.693}{T_{1/2}} = \frac{0.693}{1.6 \text{ s}} = 0.43 \text{ s}^{-1}$$

The 1.6×10^{17} atoms in the sample can produce up to 1.6×10^{17} decays, so activity = λN
    = $(0.43 \text{ s}^{-1})(1.6 \times 10^{17} \text{ decays}) =$
    6.9×10^{16} decays/s = $\boxed{6.9 \times 10^{16} \text{ Bq}}$.

22 Subatomic Physics

NUCLEAR REACTIONS

1. c   **5.** c
2. a   **6.** d
3. b   **7.** b
4. d   **8.** c

9. Uranium-238 has a tendency to absorb neutrons, so that any fission reaction that takes place in uranium-238 loses energy. This prevents a chain reaction from occurring. Uranium-235 produces enough neutrons and does not reabsorb them in the process for a chain reaction to occur, and so sustained energy from fission is produced.

10. For fusion of hydrogen isotopes to occur, separate nuclei must be brought close enough together for the strong force to take effect and fuse the nuclei together. This means that the electrostatic repulsion between the nuclei must be overcome by forces even greater than that repulsion. Fusion reactors have as yet not been able to sustain such forces long enough for practical use of the energy obtained from the fusion reactions. Fission reactors do not require anything more than fuel that can produce chain reactions, and substances that can control those reactions, so these reactors have proven much more practical to construct and operate.

22 Subatomic Physics

PARTICLE PHYSICS

1. a
2. c
3. b
4. c

5. b
6. c
7. d
8. a

9. Both hadrons and leptons are the fundamental particles that constitute all matter. Leptons, of which electrons and neutrinos are the most important constituents, are fundamental particles that are not made up of smaller particles. Leptons participate in the electromagnetic, weak, and gravitational interactions, but not the strong interaction. By contrast, hadrons are made up of different combinations of quarks, and hadrons participate in all four fundamental interactions.

10. The charge for the u quark is $+\frac{2}{3}e$ and the charge for the $\bar{d}$ quark is $+\frac{1}{3}e$. The total charge of the pion is $+\frac{2}{3}e +\frac{1}{3}e$, or $+e$.